Classic Geology in Europe

Iceland

CLASSIC GEOLOGY IN EUROPE

1. *Italian Volcanoes* Chris Kilburn & Bill McGuire
   ISBN: 9781903544044 (2001)
2. *Auvergne* Peter Cattermole
   ISBN: 9781903544051 (2001) *Out of Print*
3. *Iceland* Thor Thordarson & Ármann Höskuldsson
   ISBN: 9781780460925 (Third edition 2022)
4. *Canary Islands*
   ISBN: 9781903544075 (2002) *Out of Print*
5. *The north of Ireland* Paul Lyle
   ISBN: 9781903544082 (2003) *Out of Print*
6. *Leinster* Chris Stillman & George Sevastopulo
   ISBN: 9781903544136 (2005) *Out of Print*
7. *Cyprus* Stephen Edwards et al.
   ISBN: 9781903544150 (2005)
9. *The Northwest Highlands of Scotland* Con Gillen
   ISBN: 9781780460406 (2019)
10. *The Hebrides* Con Gillen
    ISBN: 9781780460413 (forthcoming 2022)
11. *The Gulf of Corinth* Mike Leeder et al.
    ISBN: 9781903544235 (2009) *Out of Print*
12. *Almeria* Adrian Harvey & Anne Mather
    ISBN: 9781780460376 (2015)
13. *The Southern Pennines* John Collinson & Brian Roy Rosen
    ISBN: 9781780461007 (forthcoming 2022)

For details of these and other earth sciences titles from Dunedin Academic Press see www.dunedinacademicpress.co.uk

Classic Geology in Europe 3

Iceland

Third Edition

Thor Thordarson
Faculty of Earth Sciences, University of Iceland

Ármann Höskuldsson
Institute of Earth Sciences, University of Iceland

EDINBURGH ◆ LONDON

Published by
Dunedin Academic Press Ltd
Head Office: Hudson House, 8 Albany Street,
Edinburgh, EH1 3QB
London Office: 352 Cromwell Tower, Barbican,
London, EC2Y 8NB

ISBN:
9781780460925 (paperback)
9781780466460 (epub)
9781780466484 (PDF)

MIX
Paper from
responsible sources
FSC® C167221

First edition published 2002 by Terra Publishing
(now part of Dunedin Academic Press)
Second edition 2014, reprinted 2015, 2017 and 2019
This Third edition published 2022

British Library Cataloguing in Publication Data
A catalogue record for this book is available from the British Library

Typeset by Biblichor Ltd, Edinburgh
Printed in Poland by Hussar Books

Contents

Preface to the Third Edition

Iceland is a 'wonderland' when it comes to earth sciences, and in a class with other geological wonders on Earth. Mother Earth is very much alive in Iceland, continually working the land, and in doing so, keeping its inhabitants on notice regarding the might of the natural forces. It is an integral part of the Icelandic psyche and essential for a full appreciation of the country's history. The purpose of this book is to present the basics of Icelandic geology and its influence on Icelandic culture.

In writing a book such as this, authors rely on innumerable written sources representing countless hours of work by a large group of researchers. However, it is impractical to include hundreds of direct citations to individual works in a text of this nature, as it would make it very difficult for the introductory reader. Nevertheless, we would like to extend our gratitude to the scientists whose work has made the writing of this book possible, and we dedicate it to them. Some of these contributions are listed in the Bibliography, and acknowledgements for illustrations borrowed with permission from other texts are listed overleaf.

We have kept the general framework of the second edition. In addition to some revision and correction of the original text, we have, in Chapter 10, added a new section on the 2014–15 eruption at Holuhraun, along with adding and updating the figures and tables where appropriate. One of the main changes in this edition is that regional geological maps are now superimposed on a digital elevation model, providing a 3D perspective to those figures. Detailed bathymetry has been added to the map of the Vestmannaeyjar archipelago (Chapter 4). Finally, we thank Anthony Kinahan and all those working with Dunedin Academic Press for their patience and support in the making of this third edition.

Thor Thordarson
Ármann Höskuldsson
March 2020

Publisher's Note:

This third edition of Iceland was ready for press in March 2021 when the 2021 Fagradalsfjall eruption occurred. As the Covid-19 Pandemic was then preventing students from travelling on field trips to Iceland, the authors agreed to delay publication so that they could include some coverage of 2021's eruption in the new edition. As international travel resumes, it is evident that there will be much to learn from the 2021 Fagradalsfjall eruption, which is already the longest-lasting eruption in Iceland in the 21st century. However, at this early stage, information is incomplete. A brief section on the 2021 Fagradalsfjall eruption has been added to the book at page 239 as an Addendum.

Acknowledgements

The authors would like to acknowledge the following sources for permission to use copyright material in the preparation of the illustrations, all of which have been modified. The full bibliographic details are in the Bibliography (pp. 248–250). All other illustrations are the work of the authors themselves.

Figure Source

1.1	Saunders et al. (1997)
1.2	Sæmundsson (1979)
1.3	Jóhannesson (1980)
1.4	Steinþórsson (1981)
1.5	Jóhannesson & Sæmundsson (1998)
1.6	Guðmundsson (1995)
1.7a	Alfreð Jónasson, courtesy of Sólarfilma
1.7c	Marta Bergman & Bergsteinn Gizurarson
1.9d	Self et al. (1998)
1.10b	Lockwood & Lipman (1980)
1.10d	McPhie et al. (1993)
1.11	Thordarson and Höskuldson (2008)
1.13c	Ágúst Guðmundsson
1.15b	Ágúst Guðmundsson
1.15c	John Maclennan
1.16	Sæmundsson (1979)
1.17a	Einarsson (1991)
1.20	Oddur Sigurðsson
1.21	Geirsdóttir & Eiríksson (1994a)
1.22	Walker (1966)
2.1	Jóhannesson & Sæmundsson (1998)
2.2	Geirsdóttir & Eiríksson (1994b)
2.3	Geirsdóttir & Eiríksson (1994b)
2.6	Kjartansson (1973)
2.7c	Sigurgeirsson (1995)
2.8	Jónsson (1978)
3.1a	Jóhannesson & Sæmundsson (1998)
3.1b	Einarsson & Björnsson (1979)
3.2	Einarsson (1991)
3.3	Sæmundsson (1992)
3.4b	Einarsson (1991)
3.7	Larsen & Thorarinsson (1977) and Larsen et al. (1999)
4.1	Jakobsson (1968)

ix

ACKNOWLEDGEMENTS

5.1	Jóhannesson & Sæmundsson (1998)
5.2	Magnús T. Guðmundsson
5.4	Magnús T. Guðmundsson
5.5b	Jón Kjartan Björnsson
5.6	Steinþórsson (1966)
5.8	Oddur Sigurðsson
5.9	Larsen (2000)
5.10	Larsen (2000)
5.11	Larsen (2010)
5.12	Thordarson et al (2001)
5.13a	Larsen (2000)
5.14a,b	Thordarson & Self (2001)
5.15	Thordarson & Self (2001)
5.17	Tordarson & Larsen (2007)
5.18	Magnús T. Guðmundsson
5.19	Magnús T. Guðmundsson
5.21a,b	Kári Kristjánsson
5.21c,d	Magnús T. Guðmundsson
5.22	Thorarinsson (1958)
5.23b,c	Thorarinsson (1958)
6.1	Jóhannesson & Sæmundsson (1998)
6.2	Walker (1963) & Martin et al (2011)
6.5	Jónsson (1988)
6.7a	Blake (1966)
6.8	Oddur Sigurðson
6.9	Walker (1958)
6.10a	Walker (1963)
6.13	Walker (1962)
6.14	Walker (1962)
6.15	Walker (1966)
7.1	Jóhannesson & Sæmundsson (1998)
7.4	Einarsson (1991)
7.7	Sæmundsson (1991)
7.8	Sæmundsson (1991)
7.12	Sæmundsson (1991)
7.13	Sæmundsson (1991)
8.1	Jóhannesson & Sæmundsson (1998)
8.2	Norðdahl & Hafliðason (1992)
9.1	Jóhannesson & Sæmundsson (1998)
9.2	Jóhannesson (1980)
9.4	Björn Harðarson (1993)
9.6	Oddur Sigurðsson
9.7	Jóhannesson (1980)
9.8	Einarsson (1991)
10.3	Mohn (1877)
10.4	Hartley and Thordarson (2012)
10.6	Sigbjarnason (1996)
10.8	Gunnarsson (1987)
10.9	Larsen (1984)

Introduction

Iceland is one of the very few places where we can directly observe the Earth's growth processes in action. We all are aware of two geological fundamentals: the tangible materials that we walk on and the sometimes intangible processes that operate on our surroundings. The materials (excluding living organisms) are the rocks that make up the solid crust or the outermost shell of the Earth, and the processes represent the forces that construct and erode these crustal rocks and thus shape the landscape that we have before us. The processes that operate at or near the Earth's surface are generally categorized as external, whereas those that are at work deep below the surface, or result from such activities, are called internal.

Iceland is an exceptional natural laboratory where almost all of the principal geological processes can be observed and it is an ideal setting for laymen, as well as professionals, to witness how Mother Nature operates. Unusually high rates of volcanic activity and dynamic fault movements, coupled with rapid erosion and efficient transport and deposition of sediment, make Iceland a diverse environment where constructive processes outweigh destructive ones. Nowhere on Earth is the architecture of the spreading sea floor better exposed, where different patterns of fault lines and fault movements, along with chains of volcanoes, characterize the parts of Iceland interpreted as the boundary between the Eurasian and American plates.

Large icecaps and extensive river systems rapidly grind down the volcanic pile, dispersing vast amounts of debris from the highlands to the lowlands to form thick sequences of glacial, fluvial and lacustrine sediments. Along the seashore these formations are further modified by the pounding waves of the North Atlantic, where the often spectacular landforms exhibit an intricate balance between land construction and erosion.

Over the past seven million years Iceland has been situated at a boundary of major air and ocean masses, and consequently has been exposed to extreme climatic changes. The rocks in Iceland clearly reflect this fluctuating climate. The dynamic interplay between constructive and destructive geological processes has resulted in the juxtaposition of volcanic and

Figure 0.1 Hekla in South Iceland, the queen of icelandic volcanoes, viewed from southwest.

sedimentary rocks. Dramatic swings in the climate over the past seven million years are clearly illustrated by the occurrence of diverse rock types in alternating volcanic and sedimentary successions. These changes have also resulted in a wide range of volcanic structures, because the eruptions occurred in environments ranging from subaqueous through subaerial to subglacial (Fig. 0.1). Consequently, each rock formation reflects the flavour particular to the builders – the spectrum of internal and external processes that moulded Iceland into its current shape.

Iceland is the youngest part of a much larger volcanic province that extends from Scotland to Greenland. It is also the only part of the province that is still active and possibly the only place on Earth where the processes involved in the construction of this type of volcanic province can be observed directly. Furthermore, it provides a window into continent formation as it was early in Earth's history, because such continents are likely to have been produced by volcanism of a nature similar to that found in Iceland.

There are various ways one could organize a book of this kind, but we have chosen a format that follows an excursion that a visitor could conveniently take around Iceland. This gives us the flexibility to organize the book as a set of trips that are arranged in a rational order from the traveller's point of view. It also allows us to focus our attention on regions and localities that are popular among those who visit Iceland, as well as to point out interesting geological localities of which the reader might otherwise be unaware. Consequently, the book is by no means a comprehensive textbook on Icelandic geology, but is intended for those who want to acquire some basic information on a subject or a particular geological phenomenon. We have chosen to eliminate bibliographic references, as they would have made the text too cluttered, but a selection of further reading is provided at the end. We have also tried to restrict the use of technical terms to a minimum, but those that are necessary are explained in the Glossary and are highlighted in **bold type** on first appearance.

The first chapter provides a brief overview of the geology of Iceland and the forces that have shaped the island through its relatively short geological existence. It also includes a brief explanation of the terminology used throughout the book. In chapters 2–9 we present a set of geological excursions covering the most accessible parts of Iceland, the Lowlands. In general, the excursions are planned such that they follow or extend from Highway 1, the main artery around Iceland. We begin our excursions in Chapter 2 at the Reykjanes Peninsula, because it is the point of arrival for most travelers. The excursions in subsequent chapters are arranged in a counterclockwise fashion around the periphery of the island. The breakdown of excursions into chapters generally follows the traditional geographical subdivisions, similar to that of the 1:250 000 topographical and geological maps of Iceland. In Chapter 10 we finish up with a selection of excursions within the Interior Highlands.

The heavy emphasis on the regions closest to Reykjavík is deliberate, because many travelers spend only a few days in Iceland. Consequently, they cannot travel all around the country, but can easily make short 1–2 day trips out of Reykjavík covering the area from Snæfellsnes to Öræfajökull. However, we believe that we have included enough details from other regions to provide a geological guide to all of Iceland.

Those who are interested in more information on the geology of Iceland are encouraged to look up the references listed in the Bibliography. Geological maps of Iceland are produced by the Icelandic Institute of Natural History (https://en.ni.is/) and by Iceland Geosurvey (ISOR; isor.is/jardfraedikort-kortavefsja). Some of these maps are available in local bookstores and on the internet. Finally, all Icelandic names will appear in Icelandic (i.e. using the Icelandic alphabet); so that the reader will be able to relate the place names mentioned to maps and road signs.

Travel in Iceland

We would like to offer a few words of warning in respect of travel arrangements. Despite Iceland's location just south of the Arctic Circle, it has a relatively mild maritime climate, with a four-month-long growing season and typical July temperatures around 10–15°C. However, the northerly position of Iceland puts it at the boundary between warmer temperate and colder arctic air masses, and therefore the weather can change drastically and very quickly. It is often said that one can experience all four seasons in the space of a day, and we know from experience that this is true. Thus, proper gear is essential, as well as a healthy respect for the elements.

In Iceland the flora and fauna manage to survive at this far edge of the inhabitable environment. Consequently, vegetation is extremely fragile and easily ruined by human intervention. This is especially true of the vegetation up in the Central Highlands. In this regard we encourage responsible behaviour. Please respect the guidelines for travel provided by the authorities and use only designated paths and tracks when travelling off the main roads.

Iceland can be considered an open country because access to explore its natural phenomena is normally not restricted, but respect for private property and open land is expected. Note also that potentially hazardous areas, such as cliffs, waterfalls, volcanic craters and hot springs, are often not roped or fenced off, so be careful where you step. Furthermore, fording a stream, even in a four-wheel-drive vehicle, can prove extremely hazardous because many streams and rivers are fed by glacial meltwater. Their discharge can fluctuate wildly during the course of a single summer day, water levels being at their lowest in early morning and increasing throughout the day. As you explore and enjoy the natural wonders of Iceland, respect the elements and the environment, and remember that individuals are responsible for their own safety. As with any outdoor adventure, and especially in Iceland, extreme caution and prudence are advised.

Pronunciation

The following gives some guidance on Icelandic pronunciation.

á   h**ou**se

æ   **eye**

au   **feui**lle

ð   **th**at

ei   **day**

é   **ye**t

hv   **qu**ick

í   s**ee**n

ö   t**ur**n

ú   s**oo**n

þ   **th**ink

Last explosion of Grímsvötn eruption 2011 on 25 May at 02:00 hrs. View to west. Photo by Ármann Höskuldsson.

Chapter 1

The geology of Iceland

Introduction

Iceland is located in the North Atlantic Ocean between Greenland and Norway at 63°23′ N to 66°30′ N. It is a landmass that is part of a much larger entity situated at the junction of two large submarine physiographic structures, the Mid-Atlantic Ridge and the Greenland–Iceland–Faeroes Ridge (Fig. 1.1). As such, Iceland is a part of the **oceanic crust** forming the floor of the Atlantic Ocean. This region is known as the Iceland Basalt Plateau, which rises more than 3000 m above the surrounding sea floor and covers about 350 000 km². About 30 per cent of this area (103 000 km²) is above sea level, the remainder forming the 50–200 km-wide shelf around the island, sloping gently to depths of about 400 m before cascading into the abyss.

Iceland is geologically very young and all of its rocks were formed within the past 25 million years. The stratigraphical succession of Iceland spans two geological periods: Neogene and Quaternary (Table 1.1). The construction of Iceland is thought to have begun about 23 million years ago, but the oldest rocks exposed at the surface are 14–18 million years old. If we take the age of the Earth as one year, then Iceland was only born less than two days ago. The first regional glaciers of the Ice Age appeared in Iceland about five hours ago and only a minute has passed since the Holocene warming removed this ice cover from Iceland (Table 1.1).

The surface of Iceland has changed radically during its brief existence. The forces of nature that constantly mould and shape the face of the Earth operate faster in Iceland than in most other places. The rocks are shattered by the frequent change of frost to thaw, and the wind, seas and glaciers laboriously grind down the land. Erosion removes about a million cubic metres of land from Iceland each year, but volcanism and sedimentation

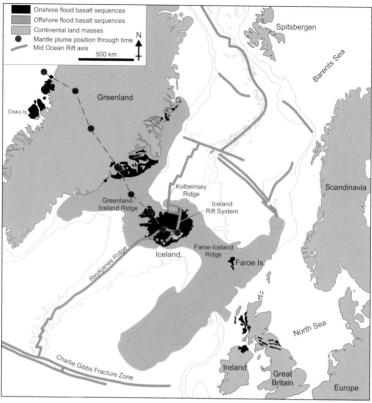

Figure 1.1 Iceland is an elevated volcanic plateau in the middle of the North Atlantic, situated at the junction between the Reykjanes and Kolbeinsey Ridge segments. The red dashed line shows the position of the Iceland mantle plume from 65 million years to the present day.

more than counterbalance this loss. The latter processes truly have the upper hand in this continuing battle, as is evident in the landmass that is now Iceland.

Geological setting: a global perspective

The prologue to the formation of Iceland includes the continental break-up that separated Newfoundland and Greenland from Europe, and the subsequent formation of the sea floor that now surrounds Iceland (Fig. 1.1). About 400 million years ago the continents on either side of the North Atlantic Ocean had merged to form a continuous landmass, which remained intact for more than 300 million years. About 70 million years ago this large continent began to break up along a fracture zone extending from the latitudes of Newfoundland–British Isles northwards into the

Table 1.1. Geologic time table for Iceland showing the terminology used in this text for geologic periods, epochs and stages. Age is shown in thousands (ky) or millions (my) years.

Era	Period	Epoch	Age	Stage	Sub-Stage	Formations	Events
Cainozoic (Cenozoic)	Quaternary	Holocene	0 3ky	Late Bog Period (sub-Atlantic)		Upper Pleistocene Formation	
			6ky	Late Birch Period (sub Boreal)			
			9ky	Early Bog Period (Atlantic)			
			10.5ky	Early Birch Period (Boreal)			
			11.7ky	Pre-Boreal			Ice Age glacier disappears
		Upper Pleistocene	12.8ky	Weichselian	Younger Dryas		Cooling, glaciers re-advance
			13.7ky 14ky		Allerød Older Dryas		Warming; Fossvogur sediments accumulate Cold snap.
			15ky 18ky		Bølling Oldest Dryas		Icelandic ice sheet retreats rapidly
			25ky 110ky				Weichselian ice sheet at greatest extension
			130ky	Eemian			Last interglacial
			230ky	Saale			Second last glacial stage
			700ky				Svinafell sediments accumulate
		Lower Pleistocene				Plio-Pleistocene Formation	Rauðsgjá tillite
							Breiðavik tillite and sediments Furuvik tillite formed Full scale glaciation
			2.5my				
	Neogene	Pliocene					Tjörnes sediments stop accumulating
			3.3my				Pacific Ocean fauna arrives in Iceland Bearing strait opens
							Tjörnes sediments begin to form
			5.3my				
		Late Miocene				Neogene Basalt Formation	First sign of cooling climate
			11.6my.				Warm temperate climate
		Middle Miocene					
			16my				Oldest rock on land
		Early Miocene					
			23my				Birth of Iceland

Arctic between Scandinavia and Greenland. As the continent on either side of the fracture drifted apart, a new plate boundary was formed.

This plate boundary is where active spreading and plate growth take place. It can be viewed as a suture where the crust is being pulled apart and molten rock (i.e. **magma**) wells up from below to fill in the gash to form new oceanic crust. As spreading continued, more magma welled up to seal the cracks and thus gradually produced the crust that forms the ocean floor around Iceland. The spreading of the sea floor is a piecemeal process and in the North Atlantic the spreading rate is about 2 cm/year (i.e. 1 cm/year in each direction). The plate boundary is delineated by series of **faults** and volcanoes, which together form a distinguishing ridge-like structure in the middle of the ocean. This structure is known as the Mid-Atlantic Ridge and its crest delineates the spreading axis. It is the suture where the American plate (to the west) and the Eurasian plate (to the east) are actively being pulled apart by the forces of plate motions (Fig. 1.1). Such **mid-ocean ridges** are found in all of the major oceans on Earth.

Normally, the mid-ocean ridges do not build up above sea level, but in Iceland they do. More magma production and eruptions than usual occur on the plate boundary across Iceland because it is a **hot spot**. The hotspot is produced by a plume of wetter, hotter and less dense material within the Earth's mantle. It is more buoyant than its surroundings and thus rises towards the surface. This structure is usually referred to as the Iceland mantle plume, and is thought to have been active for at least the past 65 million years. During this period it has brought unusual amounts of magma to the surface. As a result, a series of volcanic regions formed, which stretch across the Atlantic from Scotland to Greenland, including Iceland, which is the youngest and the only one that is still active. Collectively these volcanic regions are known as the North Atlantic **Large Igneous Province**. It is approximately 2000 km long and it represents about 10 million cubic kilometres of magma that has emerged from the Iceland mantle plume through volcanic activity over 65 million years. This figure is roughly 50 times the volume of Iceland.

Geological framework

Iceland is located at the junction between the Reykjanes Ridge in the south and the Kolbeinsey Ridge in the north. The ridges represent submarine segments of the mid-ocean ridge (i.e. the plate boundary) that is closest to Iceland. The surface expression of the plate boundary in Iceland is the

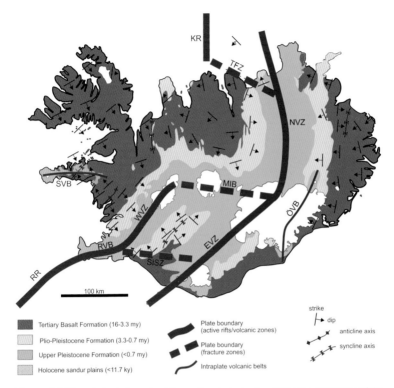

Figure 1.2 The principal elements of the geology in Iceland, outlining the distribution of the major geological subdivisions, including the main fault structures and volcanic zones and belts. RR, Reykjanes Ridge; RVB, Reykjanes Volcanic Belt; WVZ, West Volcanic Zone; MIB, Mid-Iceland Belt; EVZ, East Volcanic Zone; NVZ, North Volcanic Zone; TFZ, Tjörnes Fracture Zone; KR, Kolbeinsey Ridge (KR); ÖVB, Öræfi Volcanic Belt; and SVB, Snæfellsnes Volcanic Belt.

narrow belts of active **faulting** and volcanism extending from Reykjanes in the southwest, which zigzag across Iceland before plunging back into the depths of the Arctic Ocean of Öxarfjörður in the north (Fig. 1.2). This plate boundary is Iceland's major geological showpiece because it is the only section of the Mid-Atlantic Ridge exposed above sea level.

Above the plate boundary, the spreading rips apart the brittle crust and results in the formation of wide cracks and faults, both of which are orientated perpendicular to the spreading directions, which in Iceland are 105°E and 285°W. The spreading also results in the formation of vertical dykes that sometimes become pathways for the magma to reach the surface. Above ground, these rifts appear as swarms of linear fractures and volcanic fissures confined to narrow belts known as the volcanic zones (Fig. 1.2).

The volcanic zones are connected by large **transform faults** known as **fracture zones** or, when volcanically active, as **volcanic belts**. Together, these structures cover about one-third of Iceland ($30\,000\,km^3$); the nomenclature used for these structural identities is given in Figure 1.2.

It is important to recognize that the volcanic zones are 20–50 km-wide belts and their magma production more than matches the plate movements. Consequently, the magma that emerges at the surface through volcanic activity accumulates within the volcanic zones, more so towards the centre than to the margins. Thus, the **volcanic successions** in the centre of the rifts are buried rapidly and follow a steep path as they move away from the spreading centre (Fig. 1.3). However, the successions closer to the margins of the volcanic zones keep to much shallower paths as they travel away from the spreading axis. The excess load in the middle of the zones results in down-sagging of the crust, while concurrently the margins experience uplift, such that the successions closer to the margins are tilted, acquiring a shallow dip (5–10°) towards the spreading axis. Opposite to the situation on mid-ocean ridges, this tilt remains with the rock pile as it drifts out of the volcanic zones and is accreted to the plates on either side. Thus, together, the spreading and the volcanism produce a shallow **syncline**, which explains why the regional dip of the flanking older successions is generally towards the currently active volcanic zones (Figs 1.2, 1.3).

As mentioned earlier, the construction of Iceland resulted from interaction between a spreading plate boundary and a more deeply rooted mantle plume. The overall morphology and geological architecture of Iceland is a representation of this interaction and its development through time. The most obvious manifestation of the interaction is the elevation of Iceland over the surrounding sea floor. Here the buoyant mantle plume has pushed up the crust to form a large oval bulge. The submarine Faeroe–Iceland and Greenland–Iceland ridges on either side of Iceland are older traces of this bulge that have been displaced from their original position by seafloor spreading.

The centre of the Icelandic mantle plume is below the northwestern part of Vatnajökull icecap (Fig. 1.1). It is modelled as a 200–300 km-wide cylindrical zone of highly viscous semi-solid material that is buoyant because it is hotter and wetter than its surroundings, rising extremely slowly from depths of 400–700 km. As it approaches the surface, parts of the plume melt and thus provide magma to the volcanoes. About 23 million years ago

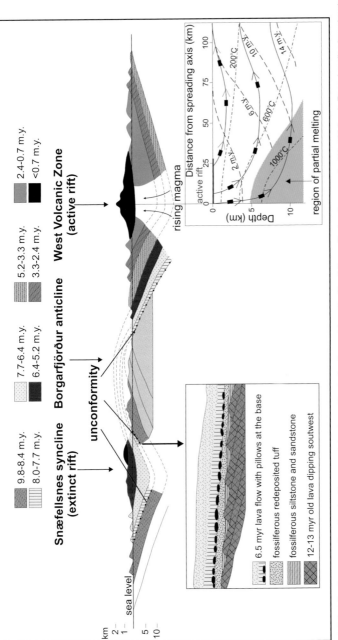

Figure 1.3 A stylized cross section showing the general structure of the Icelandic crust from the Snæfellsnes Peninsula across the West Volcanic Zone. Loading by volcanism results in down-sagging of the crust at point of greatest loading and an uplift along the margins because of displacement of ductile mantle material. This process tilts the strata towards the volcanic zones, forming a shallow syncline centred on the spreading axes and a shallow anticline in the region between the volcanic zones. The formation of volcanic zone by propagation through older crust results in formation of major unconformities as is illustrated by the inset graph on the left. The inset graph on the right shows the hypothetical spreading paths for crustal elements (black rectangles) formed within different parts of an actively spreading volcanic zone. Rocks formed near the edge of the zone follow a shallow spreading path, whereas those formed close to the spreading axis follow a much steeper path. The dashed lines are planes of equal age and the dotted lines are isotherms. The orange shaded area represents a region where the temperatures are high enough to begin to melt the subsiding rock masses.

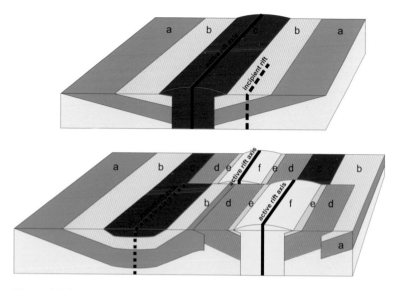

Figure 1.4 Cartoon of structural developments on a regional scale as a result of rift jumps caused by westward migration of rift zones relative to stationary mantle plume.

the centre of the mantle plume was positioned at the extrapolated intersection of the Reykjanes Ridge and the Kolbeinsey Ridge.

Although the plate-tectonic model outlined above explains the gross features of the regional geology of Iceland, it does not fully conform to all established geological facts. Distinctive reversals in dip directions in Borgarfjörður, Vatnsnes and Hreppar areas indicate broad anticlines that follow the same axial trends as the rifts. These anticlines are also associated with major unconformities and breakdown in the trends of isochrons (lines of equal age) across Iceland (Fig. 1.3). These discrepancies occur because the relative positions of the spreading axis and the mantle plume have changed with time.

If we assume that the mantle plume is a stationary structure, then the position of the spreading axis must have changed by migrating towards the west-northwest at 0.3 cm/year. This movement is in addition to the actual spreading motion that separates the plates, and it occurs concurrently. As the active spreading axis moves away, the plume responds by readjusting the position of the axis and forming new rifts closer to its centre. The rifts farther away gradually become inactive (Fig. 1.4). This process is called a rift jump.

According to this model, the currently active rifts are the West and North Volcanic Zones. The East Volcanic Zone is a rift in the making, which will eventually take over from the West Volcanic Zone (Fig. 1.2). Similarly, the precursor rift to the current West and North Volcanic Zones is a broad syncline exposed in the Neogene succession of Snæfellsnes in West Iceland and Víðidalur in North Iceland. The key consequences of rift jumps are the trapping of older crust between the two rift segments and the overlap in activity on the established versus the developing rift segments with respect to time. The accumulation of volcanic material within the rift zones results in sagging of the crust at the point of maximum loading, while the margins of the zones are uplifted. This results in tilting of the adjacent rock piles towards the volcanic zones. Consequently, the trapped crustal segment acquires the form of a broad anticline (Fig. 1.3). Thus, shifts in the position of the spreading axis explain the occurrence of synclines and anticlines in the geological succession of Iceland.

There are two active **intraplate volcanic belts** in Iceland where young (< 2 million years old) volcanic rocks rest unconformably on older formations, indicating a significant interval in volcanic activity and outpouring of magma. The first is the Öræfi Volcanic Belt, to the east of the plume centre and the current plate margins. Second is the Snæfellsnes Volcanic Belt in West Iceland, which sits on the postulated mantle-plume trail and is in part superimposed on an extinct volcanic zone, precursor to the West Volcanic Zone.

The Öræfi Volcanic Belt probably represents an embryonic rift. It is possible that the westward migration of the entire Iceland rift system is forcing the plume to establish a new volcanic zone by melting its way through the crust of East Iceland. Thus, the Öræfi Volcanic Belt may indicate yet another jump in the spreading axis across Iceland.

Volcanism

Iceland is well known for its volcanic activity. Questions often asked are:

- How many eruptions have taken place in Iceland?
- What is the number of active volcanoes?

The answers are not as straightforward as might be expected. The definitions of a volcano and an eruption encompass a wide range of phenomena that deviate significantly from the simple image of a conical mountain with smoke rising from its top.

It is not easy to say how many eruptions have occurred in Iceland throughout its geological history because it is not always obvious what part of the rock sequence represents a single eruption. However, recent volcanic activity in Iceland shows that on average there is an eruption every three to five years. Using the more conservative estimate this translates into about 200 eruptions in the past 1000 years. With this frequency as a guide, the total number of eruptions that may have taken place since the birth of Iceland, about 23 million years ago, is in the order of 5 million.

The answer to the second question is even more difficult because it is not clear how many of the volcanoes in Iceland should be considered active. Although it is easy to count the central volcanoes that have erupted time after time in the recent geological past (22 in total; Fig. 1.5), the difficulty arises when we try to assess the status of volcanoes that erupted only once. By definition, such volcanoes are extinct, but the region in which the volcanoes are situated may still be volcanically active. To circumvent this problem, geologists have introduced the concept of a volcanic system.

A volcanic system is the principal geological structure in Iceland: a fissure swarm or a central volcano, or both, which are taken to represent surface expressions of two different types of subsurface magma-holding structures, the first a deep-seated magma reservoir, the second a shallower crustal magma chamber (Fig. 1.6). An individual volcanic system is often characterized by conspicuous tectonic architecture and distinct magma chemistry, and typically has a lifetime of 0.5–1.5 million years. Altogether there are 30 active volcanic systems on land in Iceland (Fig. 1.5).

The fissure swarms are narrow and elongated strips (5–20 km wide and 50–100 km long) of tensional cracks, normal faults and volcanic fissures (Fig. 1.6). They are thought to be the surface manifestations of elongated magma reservoirs at the base of the crust (> 8 km depth). These swarms are typically aligned sub-parallel to the axis of their host rift zone, illustrating that the fundamental force responsible for their formation is plate spreading. Wide cracks indicative of pure crustal extension are usually the most conspicuous structures on the surface. Fault scarps and **graben** are also common and indicate vertical displacement and extension of crustal blocks. Young volcanic fissures typically appear as a row of small volcanic cones, whereas fissures that erupted beneath the Ice Age glaciers occur as elongated **móberg** ridges.

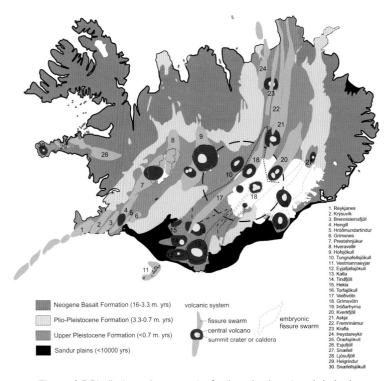

Figure 1.5 Distribution and components of active volcanic systems in Iceland.

Neogene Basalt Formation (16-3.3 m. yrs)

Plio-Pleistocene Formation (3.3-0.7 m. yrs)

Upper Pleistocene Formation (<0.7 m. yrs)

Sandur plains (<10000 yrs)

volcanic system

fissure swarm

central volcano

summit crater or caldera

embryonic
fissure swarm

1. Reykjanes
2. Krýsuvík
3. Brennisteinsfjöll
4. Hengill
5. Hróðmundartindur
6. Grímsnes
7. Prestahnjúkur
8. Hveravellir
9. Hofsjökull
10. Tungnafellsjökull
11. Vestmannaeyjar
12. Eyjafjallajökull
13. Katla
14. Tindfjöll
15. Hekla
16. Torfajökull
17. Veiðivötn
18. Grímsvötn
19. Þórðarhyrna
20. Kverkfjöll
21. Askja
22. Fremrinámur
23. Krafla
24. Þeystareykir
25. Öræfajökull
26. Esjufjöll
27. Snæfell
28. Ljósufjöll
29. Helgrindur
30. Snæfellsjökull

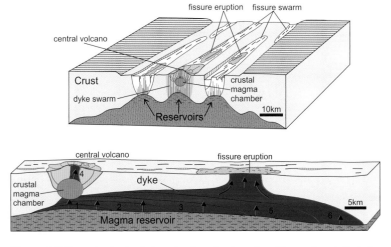

Figure 1.6 The main structural elements of a volcanic system. The numbers indicate the growth sequence of a dyke rising through the crust in a major eruption episode. Upper panel, cross section; lower panel, long section.

When present within a volcanic system, the central volcano is the focal point of eruptive activity. It is the manifestation of a crustal magma chamber that typically is situated 2–6 km below the surface. Usually, the central volcanoes are the largest volcanic edifice within each system and are often capped by a collapsed crater called a caldera (Figure 1.6).

Events on the volcanic systems are intimately linked to the plate movements. The spreading and subsequent rifting of the crust that take place at the plate boundary are not continuous in either time or space. They occur as distinct rifting episodes confined to a volcanic system at any one time. Normally, the whole system is activated in these episodes, which typically last between several years and decades. Recurring earthquake swarms and volcanic eruptions on the fissure swarm and within the central volcano characterize these rifting episodes. It is traditional in Iceland to refer to such eruptions as eldar, which translates as '**fires**' (e.g. the Krafla Fires).

Magma types

Magmas are fluids that are formed by melting of rocks in the upper mantle or the lower crust at 700–1300 °C. Melting at high temperatures (> 1100 °C) produces mafic magmas that are relatively low in silica (< 52%) and rich in magnesium, iron and calcium. On the other hand, felsic magmas are formed by melting at relatively low temperatures (700–900 °C) and are rich in silica (> 64%), but also contain significant amounts of aluminium, sodium and potassium (Table 1.2). Magmas of intermediate compositions (< 52–64% SiO_2) are formed by melting at intermediate temperatures or by mixing of mafic and felsic magmas.

Because temperature increases steadily with depth below the surface, mafic magmas generally formed at greater depths than felsic magmas. Consequently, the deeper-seated magma reservoirs (Fig. 1.6) contain mafic magma that originated by **partial melting** of the mantle and consequently modified in these reservoirs via cooling, crystallization and mixing before erupting at the surface. Magma rising from the reservoirs is the main source of volcanic eruptions on the fissure swarms, which accounts for the dominance of basaltic volcanism in Iceland.

The story is very different for the magma chambers located in the crust below the central volcanoes, because they may contain all three magma types. Mafic magma accumulates in the magma chambers from injection below, whereas felsic magma is produced by partial melting of the

Table 1.2. Classification of magma types based on chemical composition and nomenclature of igneous rocks with reference to environment of solidification.

Environment of emplacement	Geologic terms		Mafic (<52% SiO$_2$)	Intermediate (52-64% SiO$_2$)	Felsic (>64% SiO$_2$)
Surface	Extrusive	Volcanic	basalt	andesite	dacite/rhyolite
Upper crust	Intrusive	Hypabyssal	dolerite (diabase)	micro-diorite	micro-granite
Lower crust	Intrusive	Plutonic	gabbro	diorite	granite (granophyre)
Colour			dark	Intermediate	Light

surrounding crustal rocks. Mixing of the other two magma types forms intermediate magmas, induced by heating from the hotter mafic magmas. Magma chambers are the source of eruptions at central volcanoes, which explains why they are the only volcanoes in Iceland that erupt intermediate and silicic magmas in addition to basaltic magma.

The magma rises from depth towards the surface through conduits contained within the volcanic systems. In the process it cools and crystallizes to form igneous rocks. Magmas that are injected into the Earth's crust, and cool and crystallize below the surface, are called intrusive rocks. Those that become trapped in the deeper and hotter part of the crust cool slowly to form coarsely crystalline bodies, known as plutonic rocks. In Iceland such rocks occur in the root zone of central volcanoes and represent solidified magma chambers. They are grouped according to composition into gabbro, diorite and granite.

Closer to the surface the temperature is lower and the magma cools a little faster, resulting in the formation of medium to fine-grain hypabyssal rocks. Intrusive rocks of this type typically occur as sills and dykes, and are named dolerite, microdiorite, and microgranite in accordance with their chemical make-up. Sills are aligned horizontally or parallel to the stratification of the surrounding rocks. Dykes are vertical structures that cut across the stratification and often represent feeder conduits to eruptions.

Magma erupted onto the Earth's surface cools very rapidly to form fine-grain to very fine-grain extrusive or volcanic rocks, grouped as basalt, andesite and rhyolite, in accordance with their chemical composition.

Rocks

As different types of cells unite in a variety of combinations to form different organisms, **minerals** combine in innumerable ways to form a perplexing variety of rock types. Consequently, it is the mineral types and their relative abundance that determine the characteristic texture and the bulk chemical composition of a rock, just as the network of cells controls the shape and make-up of living organisms. In the same way as we divide living organisms into the animal and plant kingdoms, rocks can be grouped on the basis of their origin and overall characteristics into three families: **igneous, sedimentary**, and **metamorphic**. Igneous (fire-formed) rocks are formed by solidification of magma as the result of cooling, whereas sedimentary (settling) rocks are formed by induration of sediment by cementation or other processes at or near the Earth's surface.

Metamorphic ('change of form') rocks are formed by re-crystallization of pre-existing solidified igneous or sedimentary rocks as they are buried in the Earth's crust and are re-introduced to high temperatures or high pressures, or both. Metamorphic rocks are absent from Iceland's stratigraphical succession, but are found as ice-rafted boulders on beaches.

The Earth's crust was initially formed from magma and about 95 per cent of it is igneous rocks or their metamorphosed equivalent. Sedimentary rocks are confined to the outermost skin of the crust (i.e. the top 2–3 km). On the continents about 75 per cent of the rocks we see are sedimentary, mainly because the external processes that operate relentlessly to produce them are confined to the Earth's surface. The internal processes are mostly hidden from our view, and so are the bulk of the igneous and metamorphic rocks that they produce.

In Iceland the situation is reversed, because the majority (75–90%) of the rocks exposed on the surface are igneous, and as such they reflect well the **lithologies** that make up the Earth's crust. About 90 per cent of the igneous rocks in Iceland are mafic; the rest are intermediate or felsic. Depending on the age, sedimentary rocks account for 10% to 25% of the succession. Metamorphic rocks such as slate, schist and gneiss are not present.

Volcanic eruptions and their products

A volcanic eruption refers to expulsion of magma, gas or rocks, or all three, in any shape or form, onto the Earth's surface. Consequently, a volcanic eruption can be a single explosion that lasts only a few seconds or it can represent a continuous effusion of lava that lasts for years or even decades. An eruption can also feature a single style of activity, either purely explosive or effusive, or it can encompass both styles, because many eruptions that begin explosively transform into quiet effusion of lava over time. As a result, the definition of a volcanic eruption is not simple. It is customary among volcanologists to take any near-continuous expulsion of any combination of magma, gas and rocks, whether they are erupted in a single pulse or continually over a period of time, to be a volcanic eruption.

In essence, there are two fundamental styles of eruptions, effusive and explosive (Fig. 1.7). Effusive eruptions are characterized by a more quiet extrusion of magma, which flows away from the vent(s) as a coherent body of lava. In explosive eruptions the magma is disintegrated by the rapid expansion of gases or steam, producing rock fragments that are collectively known as tephra. Many eruptions feature both styles of activity, producing both tephra and lava, and are referred to as hybrid eruptions.

Effusive eruptions and types of lava flows

Effusive eruptions occur when magma pours out onto the surface as lava flows, and Iceland owes its existence to the steady accumulation of lava through time. Lava flows occur in various sizes and shapes, and their form is a good indicator of how readily they flowed. The fluidity of lava is controlled by its viscosity, which upon eruption is primarily determined by its chemical make-up and the temperature of the erupting magma. However, during emplacement (i.e. advance across the earth's surface) the principal control on lava viscosity is the thermal budget of the flow; that is to say, how rapidly it loses heat. It is therefore convenient to divide effusive eruptions into categories on the basis of magma composition. These categories are threefold: mafic, intermediate and felsic. At one end there are fluid mafic magmas that produce sheet-like (low aspect-ratio) lavas, often covering large areas, and at the other are highly viscous felsic magmas that form thick (high aspect-ratio) stubby flows (Fig. 1.8).

Pāhoehoe lavas (helluhraun in Icelandic) are formed on land by outpourings of very fluid mafic magma in effusive eruptions. Pillow lavas (bólstraberg in Icelandic) are formed when such eruptions occur under water. Both lava types refer to **compound lavas**, consisting of many round lobes with smooth outer surfaces.

Pāhoehoe lava flow fields in Iceland vary greatly in size, with thicknesses ranging from a few metres to several hundred metres, and lengths from 1 km to more than 140 km. Each flow is typically made up of countless lava lobes that range in size from toe-like lobes a few centimetres thick (Fig. 1.9a) to sheet-like lobes tens of metres thick.

Figure 1.7 (a) Effusive eruption, (b) explosive eruption, (c) hybrid eruption.

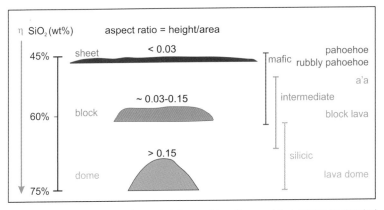

Figure 1.8 Nomenclature and geomteric forms of lava flows and their relation to chemical composition (i.e. silica content) and lava viscosity.

Figure 1.9(a) Pāhoehoe lava.

Similarly, pillow lavas consist of many bulbous sack- or pillow-shaped lobes that in long section are often tube-like (Fig. 1.9b). The space between pillow lobes is often filled with ash-size debris consisting of angular glass fragments formed by disintegration of the pillow surface as a result of quenching in water.

Figure 1.9(b) Pillow lava.

Figure 1.9(c) Rubbly pāhoehoe.

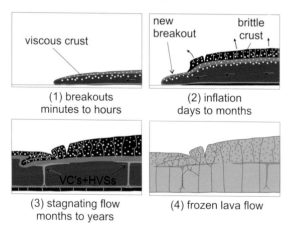

Figure 1.9(d) Snapshots showing growth of a pāhoehoe lobe by inflation from initial breakout at the active flow front (1) to last stages of lava emplacement (3). Arrangements of internal flow structures in a solidified lava lobe are shown in (4).

In pāhoehoe and pillow, the lava is transported from the vents in lava tubes beneath a coherent stationary crust. The crust is highly insulating and it minimizes cooling of the lavas during transport to the active flow front. As the lava breaks out of the tube, it forms a new lobe that is immediately sealed with new crust. The lobe grows in thickness and length by continuous injection of lava into its molten interior. Eventually, cooling increases the stiffness of the leading edge of the lobe and this slows down its advance. Gradually the flow of lava through the lobe becomes confined to the new extension of the tube in the core of the lobe. A new lobe is formed at the front of the extended tube and the process described above is repeated. This mode of emplacement is known as lava inflation (Fig. 1.9d). It is by this step-wise mechanism of forming lobe after lobe that pāhoehoe lavas gradually make their way across the landscape without much cooling of the flow interior, and in doing so they have formed some of the largest lava flows on Earth.

'Ā'a lavas (apalhraun in Icelandic) are produced by emission of more viscous mafic and intermediate magmas in mixed eruptions. They are generally thicker (3–40 m) and shorter (1–15 km) than pāhoehoe flows. 'Ā'a lavas are characterized by steep jagged flow fronts and exceedingly rough clinker surfaces formed by tearing of viscous incandescent lava during flow (Fig. 1.10a). The advance of 'a'a typically occurs in broad open channels, where the lava can flow very quickly (up to 70 km/hr). The advance of the flow fronts is much slower (< 2 km/hr) because the lava fans out and slows down as it emerges from the channels. The flow front creeps forwards and steepens until part of it becomes unstable and breaks off. Such collapses happen over and over again as the lava moves slowly forwards over its own clinker. Internally, 'a'a flows consist of coherent lava with highly irregular outlines sandwiched between layers of clinker (Fig. 1.10b).

Pāhoehoe and 'a'a lavas often occur together in the same flow field and a down-flow transition from the former to the latter is commonly observed. This type of pāhoehoe to 'a'a transition generally results from an increase in viscosity caused by cooling of the lava during emplacement. Enhanced cooling is most commonly brought about by the break-up of the upper lava crust. The majority of basaltic lava flows in Iceland have a rubbly flow top (i.e. resembling 'a'a) and a smooth basal surface like that of pāhoehoe lavas, thus representing lavas that are transitional in character (Fig. 1.11). These lavas categorically fall between pāhoehoe and 'a'a lavas and have been named 'rubbly pāhoehoe' (Fig. 1.9c).

(a)

(b)

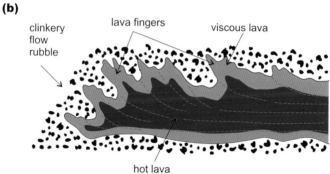

clinkery flow rubble

lava fingers

viscous lava

hot lava

(c)

(d)

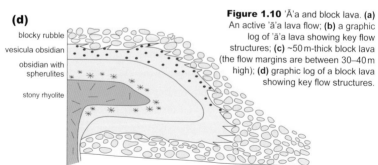

blocky rubble

vesicula obsidian

obsidian with spherulites

stony rhyolite

Figure 1.10 'Ā'a and block lava. **(a)** An active 'ā'a lava flow; **(b)** a graphic log of 'ā'a lava showing key flow structures; **(c)** ~50 m-thick block lava (the flow margins are between 30–40 m high); **(d)** graphic log of a block lava showing key flow structures.

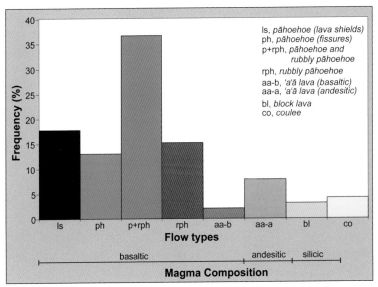

Figure 1.11 Frequency of lava flow types in Iceland (see key for details).

Block lavas and lava domes are generated by effusive eruption of felsic magmas. They tend to be thick and short lava flows, typically 0.5–3 km long and 40–800 m thick (Fig. 1.10c). The core of these lavas typically consists of flow-banded **stony rhyolite** and is surrounded by a layer of black vitreous **obsidian**, which may contain **spherulites**. The outermost shell of the lava consists of thick rubble made up of large blocky lava fragments that often reach a diameter of 1 m or more (Fig. 1.10d). When the magma is extruded from the vent, the outer surface cools and solidifies. As a result of continued extrusion of viscous magma into the molten interior, the lava swells, fragmenting the cooler and near-solid lava margins in the process, forming the blocky rubble.

Explosive eruptions and their deposits
Explosive eruptions occur when molten magma is fragmented by extremely violent boiling in the upper part of the volcanic conduit. If the explosive fragmentation is solely attributable to the expansion of dissolved gases escaping from the boiling magma, the eruption is said to be magmatic. The power (or explosiveness) of a magmatic eruption is determined by the magma's viscosity and the amount of gas dissolved in the magma at depth. Generally, felsic magmas produce the most powerful explosions because they have the

highest viscosity and gas content – a combination that allows for the greatest build-up of internal pressures in the bubbling magma. Mafic eruptions are usually the least explosive because the viscosity and gas content of the magma is low. The bubbles formed expand and escape without significant pressure build-up and the resulting activity is weakly explosive.

If the fragmentation of the magma results from explosive interaction with surface water, groundwater or glacial meltwater, the eruptions are referred to as hydromagmatic. This process can be quenching-induced granulation of the particulate matter in an erupting gas–particle jet passing through a body of surface water (i.e. sea, lake or water in an ice cauldron). In this case the water to magma interaction is passive in that it does not add any explosive energy to the system, only changes the grain size distribution of the erupted particulate matter. This is the most common style of hydro-magmatic activity in Iceland as demonstrated by the highly vesicular (fully expanded: >75% vesicularity) nature of the particulate matter produced by these events. In some cases, hydromagmatic explosions are generated by the instantaneous transformation of water into steam upon direct contact with the much hotter magma. The explosive power of these eruptions pri-marily depends on the water:magma mass ratio, and the optimum mass ratio is ~0.3. This style of hydromagmatic activity primarily produces sec-ondary eruptions that take place when lava flows over water-logged ground and form the notorious rootless cone groups within lava flow fields (see Locality 2.2; page 57).

Tephra or **pyroclasts** are formed when molten magma is fragmented by explosions, and the products of such eruptions are collectively known as **volcaniclastic deposits** or rocks. Volcaniclastic deposits are grouped into categories of ash, lapilli, bombs and blocks. When these deposits are

Table 1.3. Grain size terminology for volcaniclastic deposits and rocks.

Grain-size	Volcaniclastic deposits	Volcaniclastic rocks	Other terms often used do indicate specific type of tephra particles	
	particle class	*rock name*	*Magmatic*	*hydromagmatic*
<0.063mm	very fine ash	very fine tuff	cuspate shards	Blocky shards
0.063–2mm	fine–coarse ash	fine–coarse tuff	Pele's tears Pele's hair	Non-vesicular blocky particles
2–64mm	lapilli tephra	lapilli tuff lapillistone	pumice scoria	Accretionary lapilli armoured clasts
>64mm	bombs (fluidal) blocks (angular)	agglomerate volcanic breccia	spatter, scoria bomb, lithic blocks	Breadcrust bombs lithic blocks

consolidated into a rock they are referred to as tuff, lapilli tuff, agglomerate or volcanic breccia (Table 1.3).

Pumice and scoria describe certain types of tephra clasts, of lapilli size or larger. Pumice is white to pale grey and has frothy (vesicularity >60%) clasts of mafic to felsic composition; it commonly floats on water. Scoria, on the other hand, are somewhat denser tephra clasts with 20–60 volume per cent vesicles. Accretionary and armoured lapilli are tephra particles that indicate hydromagmatic origin when present in volcaniclastic deposits. Accretionary lapilli are bean-like aggregates formed by the coagulation of fine ash in a moist eruption column. Armoured clasts form when a thin layer of wet ash coats larger particles such as pumice or scoria. Basaltic tephra deposits often contain drop-like particles called 'Pele's tears' and highly drawn-out threads of glass known as 'Pele's hair'; both of these are named after the legendary goddess of Hawaiian volcanoes. In Iceland it is a longstanding local custom to refer to this hair-like volcanic material as witches' hair (nornahár in Icelandic). Blobs of magma that are still molten when they hit the ground and flatten out to form disk-shaped particles are called spatter.

Explosive eruptions produce a range of volcaniclastic deposits and rocks that are categorized into two general groups, **flow** and **fall deposits**, in accordance with their mode of transport and deposition. Flow deposits are formed by laterally moving currents carrying volcaniclastic debris from a volcano. The most common flow types are pyroclastic surges, pyroclastic flows and lahars, which produce volcanic formations such as base surges, ignimbrites and **debris-flow** deposits. These currents consist of a dispersive mixture of fluid and particles. The fluid can be gas (volcanic gas, air or steam, or all three) or liquid (water, muddy water or watery mud), whereas the particles can be made up of tephra or freshly fragmented lava or older volcanic rocks, or a mixture of all three components.

Fall deposits typically form widespread but relatively thin tephra layers that drape the land and are divided into scoria, pumice, and ashfall deposits on the basis of dominant fragment type or size. Individual tephra layers tend to become progressively thicker and of coarser grain towards their source volcano. Tephra fall deposits are extremely common in Iceland, whether they occur as distinctive tephra layers in the topsoil or blanket the surroundings of the youngest volcanoes (Fig. 1.12). Some soil profiles contain more than a hundred tephra layers, a record that covers about 8400 years of volcanism. Tephra layers are very useful markers when they occur

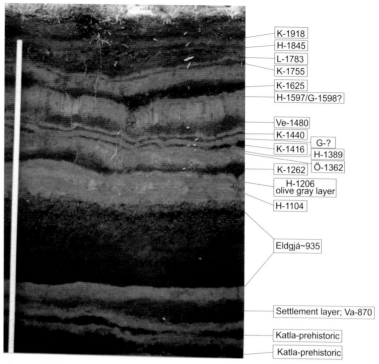

Figure 1.12 Historical tephra layers in a soil profile in southern Iceland. The text (e.g. K-1918) in the boxes indicate the source volcano and the eruption year: K, Katla; H, Hekla; G, Grímsvötn; L, Laki; Ve, Veiðivötn; Va, Vatnaöldur.

within stratigraphical successions, because their formation is instantaneous in geological terms and they are widespread. If we know the eruption dates of tephra layers, they can be used to date other geological formations as well as archaeological finds. This method is called tephrochronology and was founded in the mid-twentieth century by the Icelandic volcanologist, Sigurður Þórarinsson, through his studies of tephra layers similar to those shown on Fig. 1.12. Today, tephrochronology is widely used in volcanological studies in Iceland and elsewhere.

Volcanoes – types and architecture

A volcano is any structure produced by volcanic activity, whether it is formed by single or multiple eruptions from the same vent system. At the surface, a volcano consists of two principal components: the vent(s) and the products that issue from it. Although the overall architecture of a volcano is determined by a range of factors, the most significant ones are

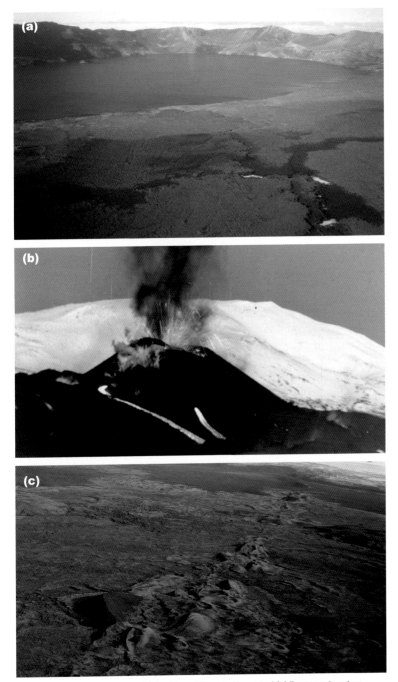

Figure 1.13 Examples of (a) caldera, (b) central vent and (c) linear vent system.

the type of magma erupted, the habitual eruption behaviour and the shape of the vent system. Consequently, volcanoes display a wide spectrum of shapes, ranging from a hole in the ground (i.e. a crater) to the majestic cone that most people commonly associate with a volcano.

The hole can be either a crater (a roughly conical pit) or a caldera, which is a large circular depression (collapse structure) surrounded by near-vertical fault-bounded walls (Fig. 1.13a). The former is created when powerful explosions excavate the ground around the vent(s). The latter is formed when a crustal magma chamber is drained rapidly during an eruption, such that its roof collapses into the void left behind by the departed magma.

In less powerful explosive, hybrid or effusive eruptions, some of the erupted material piles up around the vent(s) to form a range of structures such as ramparts, cones and even mountains. The form of these structures depends on whether the volcano was formed by eruption(s) on land, in water, or beneath a glacier, and by the shape of the erupting vent(s), which can be either circular or linear. The former shape is often referred to as a central vent, whereas the latter is called a fissure.

Central-vent eruptions typically form roughly circular cratered hills or mountains that exhibit a cone-shape geometry when formed on land (Fig. 1.13b), but a more irregular heap-like geometry when formed in sub-aqueous or subglacial settings. On the other hand, fissures are typically delineated by a row of craters or cones (or both) or by jagged ridges when such eruptions occur under water or beneath a glacier (Fig. 1.13c).

Central volcanoes

Central volcanoes are composite in the sense that they are constructed by repeated eruptions from a central vent system. They consist of multiple units of lava and tephra, ranging in composition from basalt to rhyolite. Central volcanoes in Iceland are grouped here into six categories on the basis of their average rise (i.e. slope), shape and arrangement of the main vents, and the composition of eruptives that dominate their successions (Table 1.4).

The first category includes the broad and gently sloping shield volcano, capped by fairly large (7–15 km wide) summit **caldera** and characterized by eruption of basaltic magma (Fig. 1.14a). Most of the magma is erupted as fluid lava, which through recurring eruptions builds a shield-shape volcano. The Icelandic shield volcanoes bear a strong resemblance to those of other oceanic islands (e.g. Mauna Loa and Kilauea in Hawaii), although they are

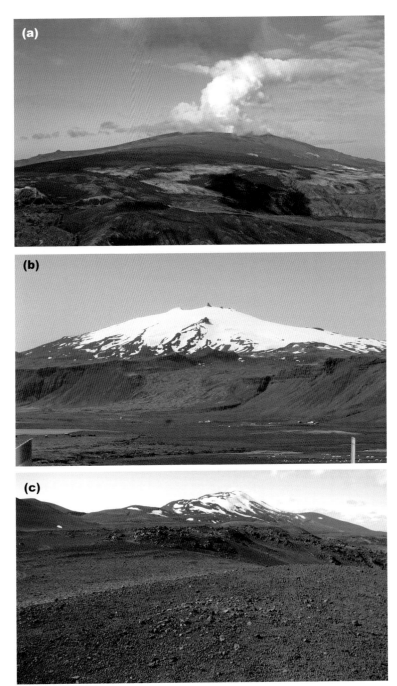

Figure 1.14 Central volcanoes: **(a)** shield volcano, **(b)** stratovolcano and **(c)** Hekla volcano.

Table 1.4 Classification scheme for composite central volcanoes in Iceland.

Environment	Type	Basal diameter	Slope	Vent shape	Magma types	Examples
Subaerial and/or subglacial	Shield volcano w/summit crater or caldera	24–38km	≤8°	circular	Basalt minor andesite or rhyolite	Hofsjökull, Eyjafjallajökull Mýrdalsjökull, Askja
subglacial (highly irregular geometry)	Móberg volcano w/summit caldera	15–26km	5–8° 20°	circular	Basalt minor andesite or rhyolite	Bárðarbunga, Grímsvötn Kverkfjöll
Subaerial and/or subglacial	Stratovolcano w/summit crater or caldera	7–23km	12–17°	circular	basalt – andesite – rhyolite	Snæfellsjökull, Snæfell Öræfajökull
subaerial	Stratovolcano ridge	6km	15°	linear	Andesite less rhyolite and basalt	Hekla
Subaerial and/or subglacial	Caldera volcano	20km	<3°	multivent complex	Basalt, minor rhyolite and dacite	Krafla
Subaerial and/or subglacial	Caldera volcano	30km	4°	multivent complex	Rhyolite, minor basalt and andesite	Torfajökull

Table 1.5. Classification scheme for monogenetic mafic volcanoes in Iceland.

Eruption characteristics		Volcano types			
Environment	Eruption type	Linear (long fissure)	Examples	Circular vent or short fissure	Examples
subaerial (magmatic)	flood lava (effusive)	mixed cone row (volume >1km³)	Laki Eldgjá	lava shield	Skjaldbreiður Kollóttadyngja
	effusive	spatter cone row (volume <1km³)	Stampar Þrengslaborgir	spatter cone	Eldborg at Mýrar Búrfell (Heiðmörk)
	effusive and explosive	scoria cone row (volume <1km³)	Seyðishólar Grábrók	scoria cone	Búðaklettur Eldfell
subaerial (hydromagmatic)	explosive	ash cone row or explosion crater row	Vatnaöldur Veiðivötn	tuff cone	Hrossaborg Hverfjall
	explosive	explosion chasm	Valagjá Kverkfjallarani	tuff ring and maar	Grænavatn, Víti ?
	effusive	pillow lava ridge	Sveifluháls Eldeyjarboði	pillow lava cone	Keilir Jólnir
subglacial and submarine	effusive – explosive	móberg ridge submarine ridge	Bláfjall	móberg cone seamount	Herðubreið, Surtsey
	effusive – explosive – effusive	table mountain	–	table mountain	

typically smaller. The second category includes the central volcanoes beneath the Vatnajökull icecap. These volcanoes have characteristics broadly similar to those of the shield volcanoes, although they exhibit a more erratic morphology because of the restrictions imposed by the surrounding glacier. These volcanoes are also crowned by a caldera as opposed to a summit crater.

The third category, stratovolcanoes, represents the classic cone-shape volcanoes with steep outer slopes. These volcanoes are capped by a summit crater or a small caldera, often lined with lava domes along the rim slopes (Table 1.4, Fig. 1.14b). They feature a diverse style of volcanism, ranging from purely effusive mafic eruptions to explosive felsic eruptions, the latter producing widespread rhyolite fall deposits and pyroclastic flows. The fourth category consists of the Hekla volcano, grouped here as a special type of stratovolcano because its vent system is distinctively linear and its products are predominantly andesite. Viewed from the southwest or northeast, the classic stratovolcano shape of Hekla is clear (Fig. 0.1). However, viewed from the northwest or southeast, it has an elongated form similar to that of a boat with its keel turned up (Fig. 1.14c). When it erupts, up to a 5 km-long fissure opens along its crest, creating a curtain of fire extending from the southwest shoulder, across the summit, to the northwest shoulder. Thus, repeated eruptions on the same fissure have built a ridge-shape stratovolcano.

In Iceland there are two active caldera volcanoes, which exemplify the last of the six categories of central volcanoes. These are Krafla, which mostly produces basalt, and Torfajökull, which mainly produces rhyolite (Table 1.4). These volcanoes are characterized by low sloping flanks and very large calderas (10–18 km wide) now filled by the products of the caldera-forming and later eruptions, featuring multiple vents inside and along the rim of the caldera. These volcanoes are dominated by effusive volcanism, although both have produced violent explosive eruptions.

Basalt volcanoes

Fissure swarms typically erupt mafic magmas and are littered with basalt volcanoes, each representing one eruption (Table 1.5). Most commonly, the eruptions take place on a fissure and form volcanoes of the linear type. Volcanoes of the central-vent type are also present and are formed by eruptions that begin on short fissures but are rapidly reduced to one vent.

Most mafic eruptions produce both lava and tephra, although not always in the same proportion. Eruptions characterized by weakly splattering lava fountains form spatter cones. More explosive Strombolian eruptions feature more vigorously spraying fire fountains, producing scoria that accumulates around the vents to form scoria cones. These volcanic cones are typically surrounded by lava dispersed from the vent either through sealed lava tubes or open lava channels.

Lava shields are a special type of volcano and are thus named because they resemble a shield (skjöldur) lying face-up on the ground (Fig. 1.15a). They consist entirely of basaltic lava and are produced by prolonged effusive eruptions from a central vent. A central crater containing a lava lake spills lava in all directions and, with time, builds up a gently sloping (<5°) and remarkably symmetrical volcano, 1–20 km in diameter and 50–750 m high. The most recent example of a lava-shield eruption in Iceland is the effusive phase of the 1963–7 Surtsey eruption, which lasted for just over three years. The largest lava shields, such as Skjaldbreiður, are likely to have been formed by eruptions that lasted for several decades and possibly centuries.

Hydromagmatic eruptions often form explosion craters surrounded by low ash cones known as tuff rings or tuff cones (Fig. 1.15b). When the explosion crater is filled with water, it is referred to as a maar (Fig. 1.15c).

Eruptions under glaciers or in the sea are common in Iceland; the 1996 subglacial eruption at Gjálp in Vatnajökull and the 1963–7 submarine eruption of Surtsey are recent examples. Similar to other basaltic eruptions, subglacial and submarine eruptions can take place on a fissure or a circular vent.

When eruptions occur through vent(s) covered by a glacier several hundred metres thick, the magma melts the ice to form a water-filled cavity at the glacier bottom and a matching depression in the glacial surface. Eventually the eruption may melt its way through the ice and form a lake. As long as there is sufficient pressure of water or ice (or both) to prevent explosive eruption, pillow lavas will accumulate around the vents (Fig. 1.16a). If the eruption stops at this stage, the resulting landform will be either a ridge or a cone made up entirely of pillow lava.

If the eruption continues, the volcano grows and the water depth will decrease. As a result, the water pressure decreases and the eruption enters an explosive hydromagmatic phase with formation of tephra that accumulates on top of the earlier-formed pillow lavas (Fig. 1.16b). The volcanic landform produced by eruptions that end at this stage is referred to as a móberg ridge or a móberg cone.

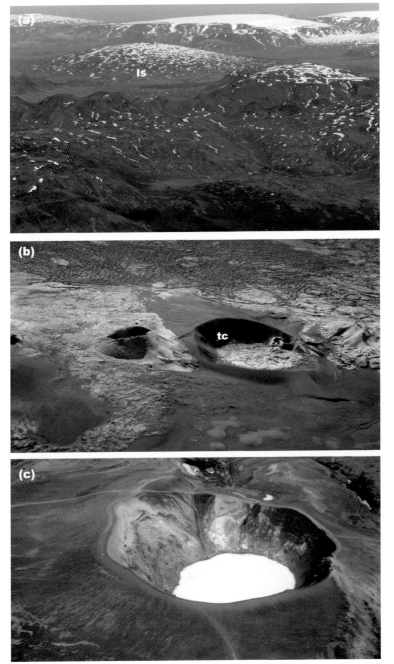

Figure 1.15 Examples of basalt volcanoes: (a) lava shield (ls), (b) tuff cone (tc), and (c) maar.

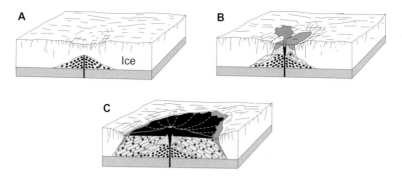

Figure 1.16 Stages of growth of a subglacial eruption. **(a)** If the eruption stops at this stage, a pillow-lava ridge is formed; **(b)** a móberg ridge forms if it stops at this stage; **(c)** the formation of a table mountain

If the eruption continues, the volcanic edifice may grow large enough to prevent water from getting into the vents, and the eruption will become purely effusive, producing subaerial lava flows. If the tephra cone is still surrounded by water, the lava simply builds upon itself a lava delta made of pillows and hyaloclastite, before continuing its advance (Fig. 1.16c). Volcanic landforms that have reached this stage form some of the most impressive volcanic structures in Iceland, known as a table mountain or a tuya (stapi in Icelandic). Submarine eruptions progress very much in the same manner as subglacial eruptions, and their volcanic architecture is identical (e.g. the Surtsey volcano, p. 107–108).

Sedimentary rocks

Sedimentary rocks are a diverse family formed entirely by external processes operating at the Earth's surface. They are divided into three basic groups: chemical, organic and clastic.

Chemical sediments are formed by precipitation from water solutions because of supersaturation of a particular chemical agent, which can be induced by cooling of the solvent (e.g. formation of silica sinter from geothermal water) or evaporation of the water, as occurs when rock salt is formed. Silica sinter formed around geothermal springs is the only naturally occurring chemical sediment in Iceland.

The chief constituents of organic sediment are the remains of organisms: shells of freshwater and marine animals are the most common (e.g. in limestone). Such deposits are not common in Iceland, but the silica shells of freshwater **diatoms** accumulate to form thick beds on lake floors (e.g. Lake Mývatn). Mass accumulation of plant debris results in the formation of sediments such as peat and lignite (surtarbrandur), the former being the main sediment fill of the many bogs in Iceland, and the latter is a primitive form of coal, which occurs as distinctive seams in sedimentary beds between the lavas of the Neogene Formation.

Clastic sediments are formed by the accumulation of debris consisting of minerals and rock fragments, which upon compaction and **lithification** become sedimentary rocks. Breakdown of existing rocks by weathering and erosion at the Earth's surface supplies the debris, which then is transported by gravity, wind, streams, glaciers or oceanic currents to be deposited elsewhere.

Similar to volcaniclastic rocks, clastic sediments are classified according to their textures. The principal classification is based on grain size – the distinguishing particle size of the sediment – and it includes familiar terms such as clay, silt, sand and breccia. When sediment is consolidated, the names change accordingly: for example, sand becomes sandstone.

Clast properties, such as angularity or roundness, are also used to distinguish between sediment types. For example, breccia consists of angular fragments that have not been modified during transport and thus represent immature sediments. Conversely, conglomerates consist of rounded pebbles, formed as angular fragments are chipped and ground down by grain collision during transport. Another important property of sediments is how well sorted the sediment is by grain size. Sediment is said to be well sorted when all of the fragments are roughly of the same size (e.g. beach sand) and poorly sorted when the sediment consists of fragments that vary greatly in size (e.g. till).

Clastic sediments are also classified according to means of transport, and take their name from the transport current or the environment of deposition: aeolian (wind-blown), glacial, fluvial, lake or marine. In these depositional environments the sediment grain size and sorting reflect the carrying capacity of the transport current, which is determined by the flow velocity and the fluid viscosity. For example, although the wind can blow hard, air has low carrying capacity because of its extremely low fluid viscosity. Therefore, aeolian deposits are normally made up of particles of sand size or finer, which typically exhibit good sorting. Water has somewhat higher fluid viscosity and therefore is capable of carrying larger particles, especially when it is rushing through ravines or during flash floods. However, as we all know, rock sinks readily in water because of its higher density. Thus, in water-dominated systems such as fluvial, lake and marine environments, flow velocity and the sediment load principally determine the grain size and sorting of the sediment. The rule of thumb is that the grain size becomes larger with increasing flow velocity and the sorting becomes less as the sediment load increases. Consequently, boulders are the first to be deposited, followed by gravel, sand, silt and clay. This is well illustrated by many river systems. Where rivers flow fast in narrow and steep channels, the bottom sediments typically consist of boulders and stones. As the slope decreases and the river channels become wider, the river slows down and smaller grains, such as gravel (pebbles) and sand, settle to the bottom. Silt and clay-size grains settle out when the flow velocity is very low, hence their common occurrence in calm lake waters and deeper parts of the ocean.

Glacial deposits differ significantly from those described above and are typically formed by mass deposition. Glaciers creep forwards as highly viscous fluid, similar to ductile tar or thick syrup, and thus are capable of carrying boulders of enormous sizes, as well as preventing the smaller particles from sinking. Therefore, glacial deposits are commonly an unstratified and chaotic mixture of clay- to boulder-size debris that is referred to collectively as till or tillite. Most glacial deposits are formed near the snout of the glacier, where its carrying capacity is greatly reduced. The basal layers of the glacier carry huge quantities of debris, which settles out to form sheet-like units of ground moraine at the glacier bottom and lateral moraines along its sides (Fig. 1.17). At the glacier snout the debris piles up to form curved end moraines, which also mark the maximum advance of the glacier. Eskers are sinuous ridges of conglomerate that are deposited by rivers flowing within the glacier. When these rivers emerge from the glacier snout, they carry large quantities of debris, which is deposited onto the broad **sandur** plain in front of the glacier.

Sediments also form as the result of mass wasting on land or in the sea. Mass wasting involves downslope movement of particles or debris under the influence of gravity without the aid of a stream, glacier or wind. The resulting sediments typically consist of unstratified and chaotic deposits containing large angular stones and boulders floating in a matrix of sand- to clay-size material (Fig. 1.18a). The architecture of these deposit types is often very similar to that of till and lahar deposits, and it is often very difficult to tell them apart. In such instances, these types of deposits are simply referred to as **diamictites**.

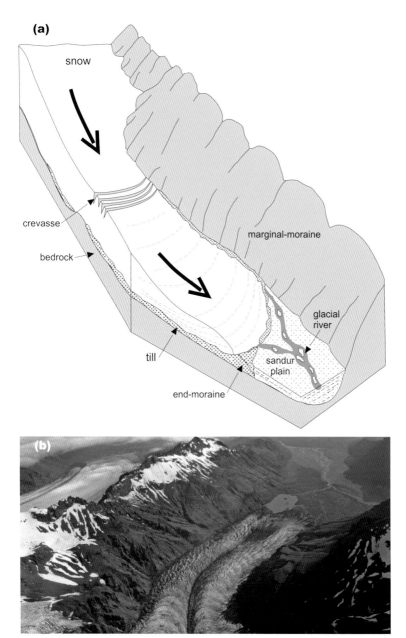

Figure 1.17 (a) Nomenclature of glacial features and glacial deposits. The transverse crevasses form above basement highs where the glacier is eroding the substrate (arrows indicate direction of flow). The till is formed as ground moraine at the glacier bottom, as marginal moraine along its sides and as terminal moraine at its front. Meltwater runs from the glaciers as rivers, forming a sandur plain in front of the glacier. **(b)** Example of an outlet or valley glacier, Skaftafellsjökull, South Iceland.

Figure 1.18 (a) Chaotic and unstratified diamictite in the Hreppar Formation in South Iceland, most probably formed by debris flow. **(b)** Rock glacier at Mt Hólshyrna above the town of Dalvík in North Iceland. Note the bowl-shape cirque above the rock glacier.

Talus slopes, an example of deposits generated by mass wasting, cover most mountain slopes in Iceland. They are formed over a long stretch of time by persistent accumulation of angular stones and blocks loosened by weathering from vertical rock faces. Another example is rock (debris) avalanches, which are common along the steep glacially carved slopes of the fjord regions of Iceland. Rock avalanches occur when large chunks of steep mountainsides or cliff faces cascade en masse into the valley below because of sudden slope failure. However, not all of the slump-like debris piles in these regions are avalanche deposits, but rather rock glaciers, formed when an icy rock mass creeps down slope with time (Fig. 1.18b).

The elements also contribute to denudation of the land by using the debris they carry as an erosional tool, and thus are an important land-shaping factor. The dust carried by the wind polishes rock outcrops and exposed boulder surfaces. Rivers use the debris they carry to deepen and widen their channels by abrasion and in the process form V-shaped valleys, gorges and canyons. Glaciers use the debris frozen into their base to grind down the land, as is evident from the striations they carve into the bedrock (Fig. 1.19). The glacial landscape is characterized by deep and steep-sided U-shaped valleys surrounded by serrated ridges and hanging valleys.

Figure 1.19 Striations marked in basalt lava by an eroding Plio-Pleistocene glacier. From Snæfellsnes West Iceland.

Geothermal activity and hot springs

Geothermal energy is important for the economy in Iceland and is mainly used for domestic heating; just over 80 per cent of the population enjoys this facility. Schools have been deliberately located near geothermal areas to utilize this source of energy for heating and swimming pools. Groundwater that percolates through cracks and voids in the top 1–2 km of the crust is heated in accordance with the temperature gradient of any particular area. In areas such as Iceland, where the temperature increases sharply with depth (50–200 °C per 1000 m), the groundwater is quickly heated to elevated temperatures (70 °C or more) and rises to the surface through fractures in the bedrock as geothermal water. The outlet vents of the geothermal water at the surface are known as hot springs (hver) and the area in which hot springs occur are known as geothermal areas. Geothermal areas are traditionally categorized on the basis of their overall average temperature: low-temperature areas, where the water is below the boiling temperature, and high-temperature areas, where the water is above it. Low-temperature geothermal areas typically occur in the regions bordering active volcanic zones, whereas the high-temperature areas are usually confined to active volcanoes. In addition, one can find cold or lukewarm carbonated springs, especially in the Snæfellsnes Volcanic Belt. These carbonated springs are sources of mineral water.

Some hot springs erupt by sending a narrow column of water and steam (and occasionally rocks) high into the air (Fig. 1.20). Such erupting hot springs are known as geysers, named after the most famous of them, the Great Geysir in Haukadalur, South Iceland. This historic geyser has amused onlookers for centuries (since 1294) with its spectacular eruptions. In the past two to three decades the Great Geysir has been very quiet and it appeared to have retired from its previous activities. However, in June 2000, earthquakes reactivated it. Other Icelandic hot springs such as Grýla in Hveragerði and Strokkur in Haukadalur have also entertained the visitors with their frequent eruptions.

Neogene, Quaternary, Holocene

The stratigraphical succession in Iceland records about 18 million years of geological history (Table 1.1), which appears trivial when compared to the 4000 million years of its nearest neighbour, Greenland. However, much can happen in 18 million years, especially in such a dynamic place as

Figure 1.20 An erupting geyser, Strokkur, the most active gusher in the Geysir geothermal field.

Iceland. Traditionally, the succession in Iceland is grouped into three major stratigraphical formations: the Neogene Basalt (18–3.3 million years), the Plio-Pleistocene (3.3–0.7 million years) and the Upper Pleistocene Formation (<0.7 million years). As we shall see below, this stratigraphical division is based on the most distinctive volcanic rock types formed during each time interval, which broadly correlate with the changes in the climate through time.

The old rocks – the Neogene Basalt Formation

The Neogene Basalt Formation represents the younger sub-period of the Tertiary and is the oldest stratigraphical formation in Iceland; it spans the time period from 18 to 3.3 million years ago (Table 1.1). It mainly appears in two large regions on either side of the active rift zones, covering about half of the country. In the east, the Neogene Basalt Formation extends from Skaftafellsfjöll in the southeast across the eastern fjords to Bakkaflói in the northeast. In the west, it stretches from Hvalfjörður in the southwest, across Snæfellsnes and the western fjords, to Bárðardalur in North Iceland (see Fig. 1.2). The cumulative thickness of the Neogene Basalt Formation is close to 10 000 m. However, the true thickness of the succession at any location does not exceed 3000 m, because the vertical accumulation of volcanic and sedimentary rocks is coupled with outward spreading of the pile.

The original Neogene landscape, which is not readily visible today, featured scattered central volcanoes, 300 m to more than 1500 m high, towering over broad and flat-lying lava plains. These plains were spotted with wetlands and dissected by the occasional gorge or river valley. Today we have an excellent cross-sectional view through the Neogene succession, because of the extensive erosion by the Ice Age glaciers, a scene rarely available within the active volcanic zones. The most distinctive outcrop feature of the Neogene rocks is its layer-cake stratigraphy, where one basaltic lava flow is stacked onto another, forming gently dipping successions hundreds to thousands of metres thick. However, a closer inspection will show that there is more to it than this.

The Neogene Basalt Formation consists mainly of volcanic rocks (> 85%) and was constructed by **volcanotectonic** processes similar to those operating in the currently active volcanic belts. Consequently, it features the same geological elements as the active volcanic zones. The exception is subglacial landforms, which are rare because the climate was considerably warmer. Thick basalt-lava series form 10–30 km-wide and 50–100 km-long lenticular bodies that represent volcanic systems. These lava piles are cut by many thin (1–10 m wide) dykes that strike parallel to the long axis of the systems. The density of these dykes is such that they often make up 3–15 per cent of the rock outcrop, and the dykes define distinctive swarms that transect the strata. In essence, the dyke swarms are the subsurface component of the fissure swarms in active volcanic systems (see Fig. 1.6). These dyke swarms are typically closely associated with clusters of andesite–rhyolite lava and tephra formations, marking the location of extinct central volcanoes. Altogether, 52 extinct volcanic systems have been identified in the Neogene succession.

Although sedimentary rocks make up only about 10–15 per cent of the Neogene succession, they are an important component because they best preserve information about the environment and the climate at the time. The most conspicuous are the red interbeds, so named because of their distinctive rusty red colour. Other types of sediments, such as finely bedded lacustrine siltstone and fluvial gravels, occur sporadically throughout the succession.

Many of the red interbeds are ancient soils that were stained red by chemical weathering in the warm and humid climate of the Neogene. Fossilized plant remains are often found in these ancient soils, which also contain fossil-rich lignite seams (i.e. primitive coal). About 50 genera of plant fossils have been found and the collection includes leaf imprints, fruit, seeds, pollen grains and tree moulds. The oldest flora, which includes trees such as swamp cypress, dawn redwood, vine and magnolia, indicates warm temperate climatic conditions in Iceland during the Miocene, similar to that of North Carolina or Portugal today. The younger Pliocene flora is characterized by conifers and broadleaf trees such as pine, spruce, alder and birch, indicating temperate climatic conditions and signifying the onset of the cooling that eventually led to the Quaternary Ice Age. The best-known Neogene fossil localities are in Northwest Iceland, including Húsavíkurkleif in Steingrímsfjörður and Surtarbrandsgil at Brjánslækur.

The middle-age rocks – the Plio-Pleistocene Formation

The Plio-Pleistocene Formation spans the time period of 3.3–0.7 million years ago (Table 1.1). It occupies a broad region on either side of the active rift zones and covers about a quarter of the surface of Iceland. On the east side it extends from Skaftártunga in the south to Langanes in the northeast. On the west side, it stretches from the Mt Esja region in the southwest to Skjálfandi in the north (see Fig. 1.2). The cumulative thickness of the formation is about 2000 m.

A large slice of the Plio-Pleistocene Formation, locally known as the Hreppar Formation, occupies the region between the active West and East Volcanic Zones in South Iceland. The formation also has three outliers, one at the Snæfellsnes Peninsula in West Iceland, another at Skagi in North Iceland (see Fig. 1.2). In these outliers, the Plio-Pleistocene volcanic products rest on glacially derived marine deposits, which in turn rest on much older rocks of the Neogene Basalt Formation. The upper surface of the Neogene succession is heavily eroded, indicating that a significant amount of time passed before the sediments covered it. Thus, here the boundary between the Neogene and the Plio-Pleistocene formations is an unconformity, because there is a gap in the stratigraphical record across it. As in the Neogene, volcanic rocks of basaltic composition are the most widespread rock type in the Plio-Pleistocene succession. Subaerial basalt-lava flows make up a significant portion of the strata, but they alternate with subglacial formations such as pillow lavas, móberg breccias and tuffs. As before, andesite and rhyolite lavas and tephras are found in significant proportions in association with extinct central volcanoes.

Towards the end of the Neogene (about 7 million years ago), the global climate had begun to deteriorate and the cooling trend that eventually led to the Quaternary Ice Age had truly set in. About 3.3 million years ago, when the first rocks of the Plio-Pleistocene Formation were forming, the 'autumn' that preceded the frigid Ice Age 'winter' had arrived in Iceland. Glaciers began to nucleate in the highlands, and the rivers grew larger because of increased precipitation. A progressive cooling resulted in incremental northward and westward growth of the Quaternary ice sheet from its nucleus in Southeast Iceland. By 2.5 million years ago it had covered more than half of the country and about 2.2 million years ago the whole of Iceland was covered by glaciers for the first time (Fig. 1.21). The Ice Age glaciation had arrived in full force.

However, this overall cooling trend was not uniform. It was characterized by frequent fluctuations between colder and warmer intervals. By 2.2 million years ago this fluctuation had developed into distinctive alternations between glacial stages, characterized by island-wide glaciation, and interglacial stages, when most of the land was free of ice. A total of nine glacial and interglacial stages have been identified in the Plio-Pleistocene Formation, indicating that each glacial/interglacial cycle lasted between 100 000 to 200 000 years.

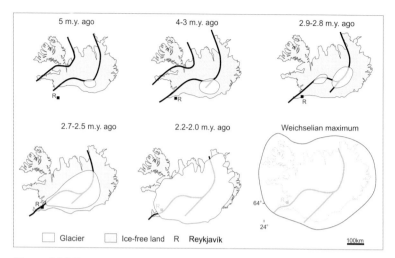

Figure 1.21 The growth of the Quaternary ice sheet in Iceland over the past five million years. The evolution of the Iceland rift system over the same period is also shown.

The changing climate had a marked effect on the landscape. Although the mechanisms of rifting and volcanism proceeded along the same general lines as in the Neogene, the face of volcanic activity changed markedly. As the glaciers grew, subglacial eruptions became more frequent. Móberg ridges and table mountains were formed in ever-increasing volume. Likewise, while rivers and glaciers grew larger, their erosive might increased, forming deeper and broader valleys. Consequently, the Plio-Pleistocene landscape was typified by irregular topography, where jagged ridges and mountains towered over steep-sided valleys. The transformation of the climate during the Plio-Pleistocene is literally written in stone, because it resulted in the formation of new rock types not found in the Neogene Basalt Formation.

During the interglacial stages, when the land was mostly ice free, the nature of the volcanic activity remained much as it did during the Neogene, forming widespread lava flows and tephra deposits. On the other hand, during the cold spells, the glaciers put down various types of till and other glacial deposits, while subglacial eruptions produced vast amounts of pillow lavas, volcaniclastic breccias and tuffs. There is a noticeable increase in the proportion of clastic and volcaniclastic sediments up through the Plio-Pleistocene Formation. In the lower part, such deposits make up 15–30 per cent of the succession, increasing to 50–60 per cent in the upper part. As the climate cooled, the red interbeds or soils become less conspicuous, while the expanding river systems were carrying forth more debris. The rivers dispersed the debris to the lowlands and to the sea, where it settled out to form fluvial, lacustrine or marine sediments in ever-increasing proportion.

The young rocks – the Upper Pleistocene Formation

The Upper Pleistocene Formation consists of rocks that are younger than 0.7 million years old, the distribution of which is for the most part confined to the active volcanic zones (see Fig. 1.2). The climate behaved in much the same way as during the latter half of the Plio-Pleistocene, with frequent alternations between colder glacial and warmer interglacial stages. Consequently, the Upper Pleistocene succession is typified by paired rock sequences, one formed under a thick glacier and the other on ice-free land. The most prominent topographic features of the volcanic zones are from this period. This includes the central volcanoes, such as Katla and Askja, and the móberg ridges and table mountains that jut through the Holocene lava cover. The most youthful-looking structures were formed during the Weichselian (most recent) glaciation, whereas more scrappy-looking structures that have suffered substantial glacial erosion date from previous glacial or interglacial stages. The younger table mountains constitute most impressive morphological structures (i.e. Herðubreið in North Iceland) and can be used to estimate the thickness of the Weichselian glacier (Fig. 1.22).

Five glacial periods have been identified in the past 0.7 million years, indicating 120 000–140 000 years' duration of each glacial/interglacial cycle. The Ice Age glacier reached its greatest extent at the maximum of the most recent glaciation, the Weichselian stage, which lasted from 120 000 to

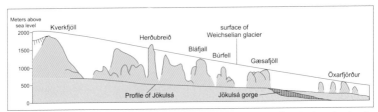

Figure 1.22 Profile from Kverkfjöll volcano to the northern coast at Öxarfjörður, showing the elevation of subglacial móberg ridges and table mountains in the North Volcanic Zone and how it is used to reconstruct the thickness of the Weichselian glacier.

10 000 years ago (Fig. 1.21). Today's glacial landscape is mainly the work of this last countrywide ice sheet.

At the maximum of the Weichselian glaciation, between 25 000 and 30 000 years ago, more than a kilometre of ice blanketed much of North America and Eurasia. The global sea level was 100–150 m lower than today, because large quantities of water were locked up as glacial ice. In Iceland, the ice sheet extended well beyond the present shores and covered the coastal shelf up to a distance of 130 km. About 20 000 years ago the bitter cold of the most recent glacial period began to decline. The global climate became milder and the Weichselian glacier began to melt. This melting released large quantities of water, and sea level began to rise. As the glacier retreated, the sea followed in its path, inundating the land, which had been pushed down by the ice load.

However, this improvement in the climate was not without setbacks, and we know of at least two periods when the cooling was great enough to see the glaciers recuperate and re-advance. These cold spells set in about 18 000, 14 000 and 12 700 years ago and are known as the Oldest, Older and Younger Dryas stages (see Table 1.1). The warm spells between them are the Bølling and the Allerød stages; times when large parts of the lowlands in Iceland were at the bottom of the sea.

The start of the Holocene is usually set at the time when the global climate became as it is today, and in Iceland this occurred during the Pre-Boreal stage, or 11 700 to 10 500 years ago, when the glacier of the Younger Dryas retreated for good (Fig. 1.21). The retreat of the Younger Dryas ice sheet was rather rapid, and as a consequence land rose extremely fast. Basically, it sprang up like a cork that has been pushed under water and then released. We know that in Iceland all of these events happened in a geological instant (i.e. less than 1200 years), because between 10 000 and 10 500 years ago recurrent explosive eruptions at the Grímsvötn volcano (five to eight events in total, referred to as G10 ka Grímsvötn eruption series, and includes the so-called Saksunarvatn tephra) produced thick tephra layers that were deposited onto ice-free land across all of Iceland, including the central highlands.

The Icelandic landscape of today was almost fully developed at the beginning of the Holocene. The exceptions are the river courses and canyons, which were formed after the melting of the glacier, and the active volcanic zones, which have been subjected to continuous modification by volcanic activity. Large parts of the sandur plains, which currently cover about 5 per cent of the land surface (5000 km²), were formed somewhat earlier by rivers emerging from the retreating glacier during the climatic transition. Since then, these plains have been modified and enlarged by the runoff from the present-day icecaps. However, this situation has changed in the last 50 years, because the outlet glaciers have retreated substantially, in doing so forming new glacial lagoons. These lagoons now capture most of the debris dislodged by the outlet glaciers. Consequently, the once braided river systems of the sandur plains are now confined to fewer and more stable river channels, which has halted the growth of the sandur.

The pattern of volcanism during the Holocene has been similar to that of the Weichselian glaciation, although its habit changed significantly. All of the currently active volcanic systems have produced eruptions in post-glacial time. Most of the magma has been erupted as lava, which today covers about 12 000 km³ of the land surface. However, a significant amount (28%) of the basalt magma came up in explosive eruptions, as is evident in the many black and grey tephra layers in the soil cover. Many of these tephra layers are formed by subglacial eruptions, which have frequently been accompanied by catastrophic floods known as **jökulhlaups**. The brown and pale tephra layers in the soils are andesite and rhyolite fall deposits dispersed by explosive eruptions from the central volcanoes.

Volcanism in postglacial times has not been evenly distributed in space or time. The most productive area has been the East Volcanic Zone in Central-South Iceland, which has the highest eruption frequency and has produced the largest eruptions. Many of the large lava shields (e.g. Skjaldbreiður and Trölladyngja) and flood lava flow fields (e.g. Þjórsá and Bárðardalur lavas) were formed in the earlier part of the Holocene, or between 5 000 and 10 000 years ago. Current data suggests that the output by lava producing eruptions in the mid- to early Holocene was considerably greater (by a factor of 2) than in the post-5000 year period.

The disappearing Weichselian glacier left behind barren and debris-covered land. The finer silt and sand fraction of the loose debris was picked up by the wind and dispersed, providing the raw material for soil formation. As the climate warmed, reaching a thermal maximum in the interval 8000 to 5500 years BP, plants began to spread out from North and Northwest Iceland, where they had survived the Weichselian glaciation on ice-free areas. Birds and oceanic currents carrying plant seeds from the surrounding continents also introduced new species. Grasses, sedges and willow were the first to appear and were followed by birch, which spread rapidly and became the dominant species.

By the time the first Norseman arrived in Iceland in the ninth century, birch forest covered about 25 per cent of the country. The vegetation changed dramatically after the arrival of human beings. Not only did it result in the introduction of new species, mainly weeds, but also in the sudden expansion of grasses and sedges at the expense of birch and willow. A few centuries after the arrival of the first settlers, the woodlands had been reduced to a fraction of what they used to be. Today woodland and scrub cover 1 per cent of the country. This sudden and severe reduction in woodlands was in part attributable to the cultivation and domestication of the land, although it was also aggravated by the cooling climate that really set in about 2000 years ago and reached climax during the Little Ice Age that began between 1250 and 1300 AD and lasted until the beginning of the twentieth century.

Aerial view of the point of Reykjanes the place where the Mid Atlantic Ridge emerges from the sea. Photo by Oddur Sigurdsson.

Chapter 2

The southwest

General overview

The geology of Southwest Iceland spans the past 3.1 million years, from the very last stages of the Pliocene through the Quaternary to the present day. The oldest rocks outcrop in and around Mt Esja in the north, the succession becoming progressively younger to the south. The youngest rocks outcrop along the axis of the Reykjanes Volcanic Belt (Fig. 2.1), which connects the submarine Reykjanes Ridge and the West Volcanic Zone. The belt consists of four volcanic systems, which are, from east to west, the Hengill, Brennisteinsfjöll, Krýsuvík and Reykjanes systems. The systems have a northeasterly strike and extend across the peninsula.

The two excursions for this region provide a general overview of the regional geology and brief descriptions of some exciting geological features at selected stops. The first excursion is centred on the greater Reykjavík area and the second one concerns the Reykjanes Peninsula.

Greater Reykjavík

A city on the margins of a mid-ocean rift

Situated on rather featureless rolling hills at the bottom of Faxaflói Bay, greater *Reykjavík* [64.1328, −21.8983] may not appear to offer much excitement to geological enthusiasts. However, there is more to it than catches the eye at first sight; at least enough to kill a day or two, if one prefers rocks to shops. Situated on the outer fringes of the Reykjanes Volcanic Belt (Fig. 2.1), it features many examples of the contrasting forces that have shaped and moulded the country in its recent geological past. These include glacial features, rift-zone volcanics and tectonics. Some outcrops are within walking distance from the city centre; others are within a

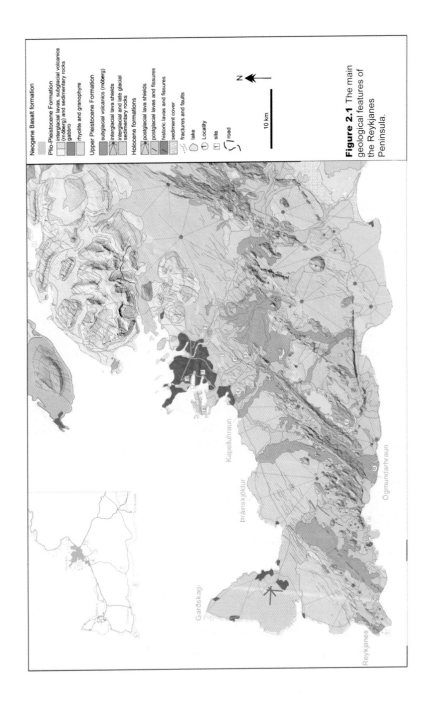

Figure 2.1 The main geological features of the Reykjanes Peninsula.

Neogene Basalt formation

Plio-Pleistocene Formation
interglacial lavas, subglacial volcanics (móberg) and sedimentary rocks
gabbro
rhyolite and granophyre

Upper Pleistocene Formation
subglacial volcanics (móberg)
interglacial lava shields
interglacial and late glacial sedimentary rocks

Holocene formations
postglacial lava shields
postglacial lavas and fissures
historic lavas and fissures
sediment cover
fractures and faults
lake
Locality
site
road

N

10 km

Garðskagi

Þráinskjöldur

Kapelluhraun

Ögmundarhraun

Reykjanes

short driving distance or can be reached by city bus, plus a hike of a few kilometres.

Reykjavík ('smoky bay') was thus named by the first settler, Ingólfur Arnarsson, in about 874 because of the columns of steam rising from geothermal springs in present-day Laugardalur. In the first decade of the twentieth century, these springs were used as washing pools, but they disappeared shortly after 1930 when their water was utilized for geothermal heating by the city of Reykjavík. The Laugardalur swimming pool, of course heated by geothermal water, is the only reminder of their past existence.

Esja [64.2406, −21.6658], the mountain that stands tall to the north of the city, is an essential component of the geological scenery of greater Reykjavík (Fig. 2.1). Esja and smaller mountains around the town of Mosfellsbær (except Mosfell) are made up of the oldest rocks outcropping in the vicinity of the capital. The Esja succession covers about 1.3 million years, the oldest rocks (3.1 million years old) exposed at the foot of Esja in the west and the youngest (1.8 million years old) near the pale-coloured *Móskarðshnjúkar* [64.2429, −21.5283] that cap the mountain in the east. The strata dip gently to the southeast; their cumulative stratigraphical thickness is about 1650 m. Eastward and upward younging of the strata, and shallowing of its dip, reflect origin by spreading from the currently active West Volcanic Zone. The bulk of the Esja succession was formed by the now-extinct Kollafjörður and Stardalur volcanic systems. However, the oldest rocks outcropping along the coast at Kjalarnes belong to yet another extinct volcano, the Hvalfjörður volcanic system. The roots of the central volcanoes that once were the loci of activity on these systems are now exposed near the base of the mountain as gabbroic intrusions. These gabbro bodies may represent the tops of a solidified crustal magma chamber. In the past these chambers supplied magma to the eruptions at the three central volcanoes exposed in the Mt Esja stratigraphical succession. Many subvertical basaltic dykes cut the Esja succession, and some of those acted as feeder conduits to fissure eruptions. Such dykes are readily visible from the main highway along the northwestern slopes of *Lokufjall* [64.2818, −21.8259] and *Eyrarfjall* [64.3193, −21.7262] (Fig. 2.1: sites 1 and 2).

The lower part of the Esja succession consists of 11 relatively thick sequences of basaltic lavas, each paired with distinctive horizons of clastic sediments or subglacial volcanic deposits, or both. Initially, it was thought that each pair represented an interglacial and a glacial period. However, recent findings show that the lowest four sedimentary horizons were formed in a glacier-free environment. They are made up of fluvial conglomerates and sandstones, often intercalated with diamictite deposits laid down by debris flows or mudflows. Three of these units are exposed in accessible outcrops at Mt Eyrarfjall, the first in a cliff exposure at sea level, the second and the third at about 100 m and 200 m above sea level (Fig. 2.1: site 2).

The remaining sediment horizons are glacial in origin. The sediments consist of tillites and associated fluvial deposits, and they are often intercalated with subglacial móberg tuffs and pillow lavas. The oldest tillite in the Mt Esja succession, and therefore in Southwest Iceland, indicates full-scale glaciation at about 2.6 million years ago. The two oldest tillite beds are exposed in outcrops at *Hvammsvík* [64.3677, −21.5679] and at *Fossá* [64.3560, −21.4559], close to Highway 1 on the south side of Hvalfjörður (Fig. 2.1: site 3).

The bedrock to most of greater Reykjavík is the so-called Grey Basalt (grágrýti in Icelandic), which is responsible for the rolling hills that characterize most of the urban landscape. The Grey Basalt is mainly covered by younger formations, but is exposed in road cuts and sporadic coastal and inland outcrops. It can also be seen in the walls of the Parliament House in downtown Reykjavík, because it is constructed from columnar blocks taken from one of the Grey Basalt lavas. At first the Grey Basalts of Reykjavík were thought to represent a single lava flow field derived from the lava shield Borgarhólar at Mosfellsheiði to the northeast of the city. Now it is known that it consists of several lava flow fields formed by sporadic eruptions during the last 500 000 years. Thin units of tillite, and fluvial and marine sediments, often separate the lava formations. When they were formed, these lavas flowed down shallow valleys, but today they cap the hills and are thus a prime example of inversion of the landscape induced by subsequent erosion, where the sediments to the side of the lava were more easily cut down by the erosion. The most widespread lava units are eroded remnants of lava shields, whereas the smaller lava units were produced by fissure eruptions.

A blanket of rather featureless till (ground moraine) laid down by the glaciers of the Weichselian period covers the bedrock. This is most apparent at higher elevations where the surface is usually littered with large boulders (urð). When the bedrock lava is exposed, its surface is usually polished and striated by glacial erosion. Sometimes such exposures exhibit the classic whaleback form of roches moutonnées. The striations and the whaleback structures are good indicators for the flow direction of the last

Ice Age glaciers. Striated lava surfaces and ground moraines formed by the main Weichselian glacier are exposed at several other localities in and around Reykjavík, and a few are indicated on Figure 2.1 (sites 4, 5, 6) [64.0480, −21.9461; 64.0475, −21.8491; 64.1471, −21.7216].

Below the peat bogs, at a depth of up to a few tens of metres, glacial deposits and glaciomarine sediments fill in the topographic depressions. Usually, at the bottom of this sediment sequence is basal till, or ground moraine, deposited by the advancing Weichselian ice sheet. The overlying end moraines, glaciofluvial and glaciomarine deposits were formed during the stage when the ice sheet was retreating from the area 10 500–15 000 years ago. The land had subsided from the burden of the Weichselian glacier, so, when it withdrew, the ocean followed in its track and submerged all of the lower-lying areas below the present-day 45 m altitude. This invasion of the ocean formed fossil-bearing deposits now exposed at several key localities (e.g. *Fossvogur* [64.1213, −21.9263]).

With the volcanic zone to the east and south of Reykjavík free of ice, the face of volcanism changed. Lava flows began to flow freely from lava shields and volcanic fissures, and several lavas made their way to the sea within the boundaries of greater Reykjavík (see Fig. 2.1).

Locality 2.1   Deglaciation history of Reykjavík – the Fossvogur sediments and other glacial deposits

The Fossvogur sediments are one of the classic geological sites in Iceland that have attracted geologists and other naturalists for more than 150 years, ever since they were first investigated in 1836 by the French naturalist Louis Eugéne Robert. These glacial and fossiliferous glaciomarine deposits tell an intriguing story, which through time has fostered widely varying ideas about their origin and age. Until recently, it was generally believed that the sediments were formed during the most recent interglacial or the Eemian stage (i.e. 110 000–130 000 years ago, Table 1.1). Now we know that the deposits are much younger and were formed during the demise of the Weichselian glacier. Radiometric ^{14}C dating shows that they were laid down between 12 800 and 15 000 years ago, during the Bølling-Allerød warm stage. Thus, the deposits are an integral part of the deglaciation history in greater Reykjavík.

In short, the deposits of the Fossvogur sequence illustrate a cyclical climatic change that began with deposition of the tillite during the Oldest Dryas stage of the Weichselian glaciation (Table 1.1). This was followed by

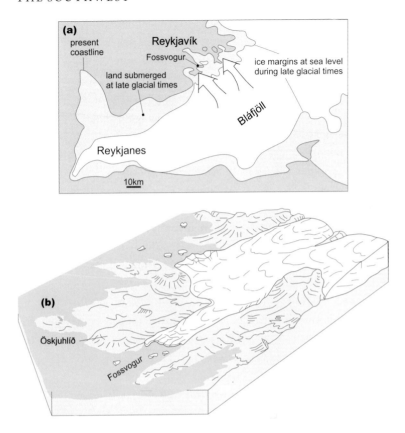

Figure 2.2 Glaciers in Southwest Iceland during the Allerød stage: **(a)** extent of the Reykjanes glacier during the Allerød stage; **(b)** snapshots of the Reykjavik area during the times of the Fossvogur tidal glacier.

deposition of fossiliferous marine sediments as a result of partial deglacia-tion and sea invasion during the warming of the Bølling-Allerød stage. Shortlived cooling in the middle of the period (i.e. during Older Dryas time) temporarily interrupted this trend and resulted in a brief advance of the Fossvogur tidewater glacier and formation of submarine debris flows (Fig. 2.2a). The sequence is capped by till and other glacial deposits formed by re-advancing glaciers during the cold snap of the Younger Dryas stage (11 700–12 800 years ago; Fig. 2.2b, Table 1.1).

The Fossvogur sediment sequence is exposed in a series of coastal out-crops, 2–5 m above the land surface, that extend for about 2 km along the northern shoreline of the inlet from which the sequence takes its name. Nautólfsvík, directly south of Reykjavík airport, provides the best beach

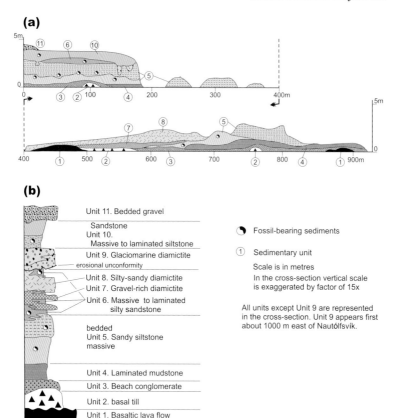

(a)

(b)

Unit 11. Bedded gravel

Sandstone
Unit 10.
Massive to laminated siltstone

Unit 9. Glaciomarine diamictite

erosional unconformity

Unit 8. Silty-sandy diamictite

Unit 7. Gravel-rich diamictite

Unit 6. Massive to laminated
silty sandstone

bedded
Unit 5. Sandy siltstone
massive

Unit 4. Laminated mudstone

Unit 3. Beach conglomerate

Unit 2. basal till

Unit 1. Basaltic lava flow

Fossil-bearing sediments

Sedimentary unit

Scale is in metres
In the cross-section vertical scale
is exaggerated by factor of 15x

All units except Unit 9 are represented
in the cross-section. Unit 9 appears first
about 1000 m east of Nautólfsvík.

Figure 2.3 (a) Coastal section showing part of the Fossvogur sediment sequence in and around Nautólfsvík; **(b)** composite graphic log of the Fossvogur sediment sequence showing the stratigraphical order of the main sediment units and their interpreted origin.

access to the Fossvogur outcrops. Those who wish to examine the outcrops from the beach are advised to do so at low tide (tidal range is 1.7–3.8 m) and put on waterproof footwear. Alternatively, one can follow the path at the top of the coastal cliffs to examine the outcrops. In outcrop the arrangement of the diamictite and fossiliferous mudstones to sandstones that make up the Fossvogur sediment sequence is rather complex (Fig. 2.3a) and its history is best described by a composite graphic log where the sediment units, labelled 1 to 11, are arranged in stratigraphical order (Fig. 2.3b).

The bedrock to the Fossvogur section is the Grey Basalt lava (unit 1) and it pokes out as whaleback humps with glacially striated surfaces. A featureless diamictite (unit 2) containing striated boulders often fills the depression between the lava humps. Together, these units represent bedrock erosion and deposition of basal till by the main advance of the Weichselian glacier.

The overlying deposits (unit 3) consist of beach conglomerates. These deposits mark the transition from a glacial to a marine environment, as the Weichselian glacier retreated and the sea began its invasion. The superimposed fossiliferous mudstone and siltstone deposits (units 4 and 5) record continued marine invasion in the early Bølling-Allerød time. The lower laminated mudstone units were formed as tidal flats. The upper massive-to-bedded sandy siltstone units and their fossilized fauna were formed in somewhat deeper water in a protected fjord environment. The abundance of **dropstones** in these units indicates that the glacier was not far away. These dropstones were carried by drifting icebergs originating from a calving glacier at the head of the fjord and dropped to the bottom of the sea as the ice melted.

Upwards unit 5 grades into fossiliferous silty sandstone (unit 6) indicating continued deposition in a submarine environment. This sequence is interrupted by beds of crudely stratified coarse-grained diamictites (units 7 and 8), which were formed by intermittent subaqueous debris and mudflows. These units contain glacially striated basalt boulders and mudstone clasts picked up by the flows through erosion of the underlying strata. This sequence is capped by a strong erosional surface and glacial deposits (unit 9) formed by the Older Dryas advance of the **tidewater glacier** into the old Fossvogur fjord. The morainal hills at the head of the current Fossvogur inlet were also formed at this time.

The fossiliferous siltstones and sandstones of unit 10 indicate renewed retreat of the glacier and sea invasion towards the end of the Bølling-Allerød stage. The bedded gravels and boulder beds (unit 11) at the top of the sequence formed when the area was once again covered with a glacier during the cold snap of the Younger Dryas stage. A raised boulder-beach deposit is found in Öskjuhlíð, the hill north of Fossvogur, at 43 m above sea level. It represents the highest stand of the sea in the wake of the Younger Dryas glacier.

To complete the history of deglaciation in the greater Reykjavík area, we must explore other sites. Fossiliferous glaciomarine sediments of similar character and identical age to the Fossvogur beds are exposed in an outcrop along the north shore of *Kópavogur inlet* [64.1051, −21.9065] (see Fig. 2.1: site 7). Near the bottom of the inlet these marine deposits are overlain by glaciofluvial sand and gravels, and an abutting end moraine that can be followed to the south up the slopes of Arnarnes. Farther out, a well-defined end moraine extends across *Álftanes* [64.1006, −22.0326] (see Fig. 2.1: site 8), which clearly represents a temporary standstill of a major glacier. This moraine was formed after the warm spell of the Bølling-Allerød stage and it represents the maximum glacial stand during the Younger Dryas stage between 11 700 and 12 800 years ago. When the Holocene warming set in for good, the glacier retreated rapidly away from these moraines, thereby leaving them as a record of its previous activities.

Locality 2.2   Rauðhólar rootless-cone group at Elliðavatn

Follow Miklubraut, the main east–west artery through Reykjavík, east towards the suburb Árbær, then turn south onto Suðurlandsvegur, the branch of Highway 1 that leads to the town of Selfoss, and follow it for about 4 km. There, on the eastern outskirts of Reykjavík just south of Highway 1 is a cluster of rather inconspicuous reddish hills that can be easily overlooked (see Fig. 2.1). These are the *Rauðhólar* [64.0929, −21.7490] ('red mounds'), which are an excellent example of rootless-cone groups, a volcanic landform that is common in Iceland as well as on the planet Mars.

Rootless-cone groups consist of closely packed cones that rest directly on the associated lava flow. Cone dimensions vary from 2 m to 40 m high and 5 m to 450 m wide. They occur in tube-fed pāhoehoe flows where the lava advanced over wetlands, such as shallow lakes or swamps. The area of individual cone groups is most commonly between 1 and 10 km^2, but the largest cone group, Landbrotshólar, covers 150 km^2. Rootless cones are formed by hydromagmatic eruptions caused by explosive interaction between molten lava and water-saturated substrate. Their cones represent volcanic vents that have lateral feeders, which are the lava tubes of pāhoehoe flows. Thus, they differ from normal vents in that their feeders are not rooted deep within the crust, hence the name 'rootless cones'.

The naturalists Sveinn Pálsson (1793) and Eugéne Robert (1840) were the first to suggest a secondary origin for the rootless-cone groups in Iceland. Another century passed before this idea was firmly established; when the Icelandic volcanologist Sigurður Þórarinsson showed that the Mývatn cone group in North Iceland was indeed formed by rootless eruptions. However, the exact mechanism of their formation was unravelled only recently by studies of the Rauðhólar cone group, and we therefore outline the results in some detail.

Rauðhólar occur as a well-confined field (1.2 km²) of small scoria and spatter cones within a 2 km-wide lava branch derived from the 5200-year-old Leitin lava shield of the Brennisteinsfjöll volcanic system. This branch of the lava extends some 27 km from its source at Bláfjöll to the sea at Elliðavogur in Reykjavík (see Fig. 2.1). Several smaller cone groups occur in the Leitin lava, including the minute but well-known Tröllabörn cone group about 5 km east of Rauðhólar. The Rauðhólar cone group is located where the lava flowed into and covered a small lake just north of present-day Lake Elliðavatn. The cone group is confined to a raised (5 m high) scoria platform that rests on 7 m of solid lava. In turn, the lava rests directly on 1–2 m-thick mudstone that covered the floor of the lake before the lava arrived.

The Rauðhólar cone group originally featured about 150 cones and craters, but roughly one third of the cone group was excavated for road ballast in the middle of the twentieth century. The largest cones are located within the northern half, which is characterized by a cluster of closely packed near-circular cones (Fig. 2.4a). The random arrangement of cones and craters is obvious when one attempts to isolate individual structures. The German volcanologist Maurice van Komorowitz described the group in 1912 and his map is the only available source that shows the original size and distribution of the Rauðhólar cones. The largest cone of the group had a basal diameter of 212 m and rose 22 m above the surrounding lava surface. The smallest cones, still seen in the southern part of the cone group, are only a few metres high and exhibit **hornito**-like forms.

The excavated part provides excellent exposure to examine the internal structure of the cone group and individual cones. These structures, which provide important clues about the formation of the cone group, are shown schematically on the reconstructed profile on Figure 2.4b. Each cone has a crudely funnel-shaped conduit (not visible) extending from the base of the flow and up through the coherent lava, terminating in a bowl-shaped crater. Another noteworthy feature is that later cones lap onto and sometimes partly conceal earlier cones. Most importantly, individual cones exhibit distinctive internal layering, showing that each cone was built from multiple explosions during a period of sustained activity.

In section, the cone ramparts typically feature a well-bedded lower sequence of ash and scoria and a crudely bedded upper sequence of

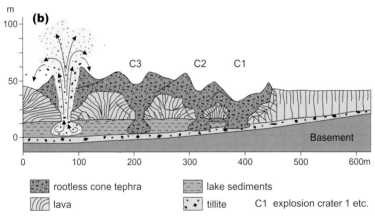

Figure 2.4 (a) Rauðhólar rootless-cone group, showing the original arrangement of pristine cones in the western sector of the group. (b) Reconstructed cross section through the Rauðhólar cone group, showing the general arrangements of cones and their characteristic internal structures.

coarser-grained **agglutinates**. The lower sequence consists of 0.2–0.6 m-thick layers composed of moderately sorted lapilli scoria alternating with thinner (<0.2 m) crudely laminated beds of red-baked lacustrine silt and black ash. The scoria layers also contain, in variable abundance, lumps of red-baked mud and clasts that have cores of either vesicular lava fragments or cooked mud armoured with a skin of quenched lava. The upper sequence features multiple 0.5–1.5 m-thick layers of brick-red agglutinates, mainly composed of spatter bombs with fluidal surfaces and variable twisted or distorted shapes. The distinctive brick-red colour of the deposit comes from a micrometre-thick coating of red-baked mud that was welded to the surface of the clasts while hot. Lapilli scoria is present in only minor amounts and the abundance of mud lumps is significantly lower. A 1–2 m-thick lava-like layer of welded spatter often caps the agglutinate beds. The upward increase in grain size within the deposits emulates the attenuation of explosive power during an individual rootless eruption and, importantly, this change goes hand in hand with a decrease in the abundance of lacustrine mud incorporated into the deposits by the explosive activity.

When considering the eruption mechanism that produced Rauðhólar and other rootless-cone groups, the following facts need to be kept in mind. First, although the cone groups rest directly on top of the host lava, their cones show no evidence of having been deformed or modified by moving lava. Secondly, the occurrence of lacustrine muds as distinct beds within the lower sequence is a clear indication of their involvement in generating steam explosions by providing the water (i.e. the coolant). Obviously the lava was the fuel. Thirdly, the internal stratification of individual cones shows that they were formed by multiple explosions of decreasing intensity, whereas overlapping arrangements of cones within the group imply a certain time sequence and duration for the explosive activity. Fourthly, we have to consider the apparently random distribution of the cones and the fact that the lava continued its advance despite the rootless eruptions. In other words, we have to get the lava across the lake bed and at the same time initiate rootless hydromagmatic eruptions through contact between hot lava and water, forming cones on top of the lava that are in no way modified by its movements.

These conditions can be met if the lava is a tube-fed pāhoehoe flow (Fig. 2.5). Initially, the lava enters the lake as relatively small pāhoehoe lobes from a set of tube entries at the active lava fronts. The insulating crust seals the lobe interiors from the water, and they then inflate and expand laterally in response to continued injection of lava. The internal

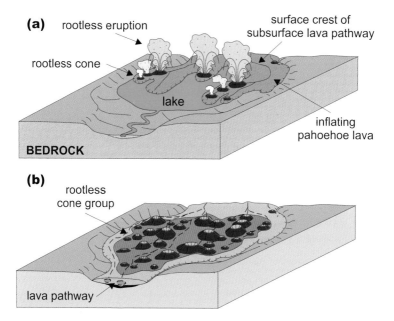

Figure 2.5 Formation of the rootless-cone groups. **(a)** The lava has covered the basin and has thickened because of lava inflation. At this stage the lava is transported across the basin in narrow lava tubes under insulating crust. Cracks in the tube floors form entries for the hot lava into the underlying water-logged lake sediments, initiating series of rootless eruptions. **(b)** A fully developed rootless-cone group.

lava tubes are thus extended to new lake entries and the process repeats itself, extending the lava farther into the lake. Another consequence of this process is that, as the lava behind the active flow fronts increases in thickness by inflation, it begins to sink into the soft mud on the lake floor. However, the subsidence is not uniform and, when cracks open in the base of the lava below the internal pathways, the glowing hot lava flows straight into the water-saturated mud and initiates the steam explosions. If the explosions are powerful enough, they burst through the overlying lava to emerge as rootless eruptions that build cones around the vents (Fig. 2.5a). At any site, the eruption stops when the supply of water (i.e. mud) or lava runs out. As the flow migrates across the lake bed, the explosive vents follow and, in doing so, gradually build a group of rootless cones on top of the lava. When the lava eventually reaches the opposite shore, it continues its advance in the same manner as it did before its encounter with the lake bed.

Locality 2.3
The Hjallar fault system and the Búrfell spatter-cone volcano

Travelling south through Rauðhólar, follow the unsurfaced road that wiggles around the eastern side of Elliðavatn, and then the signs to Heiðmörk.

The basement rocks of Heiðmörk are the Grey Basalt lavas. The lavas are partly covered by ground moraine of the Weichselian glacier, but when exposed they exhibit examples of glacially striated surfaces (e.g. site 5 on Fig. 2.1). On the eastern side of Heiðmörk, the Grey Basalt is covered by a few younger Holocene lava flows. The one that comes closest to the road is the Strípshraun flow, a basalt lava with abundant large olivine phenocrysts (>10 per cent by volume). This lava flow is 9 km long and it originated at a cone near Þríhnjúkar that has the peculiar name Eyra ('ear').

We are now within the *Hjallar* [64.0502, −21.8452] fault system, which consists of northeast-trending fissures and faults defining a series of narrow **graben** and **horst**. It marks the northern end of the fissure swarm associated with the Krýsuvík volcanic system (see Fig. 1.5). Hjallar is actually the name of a 5 km-long scarp marking the northwestern margins of the fault system, which is characterized by stepwise normal faulting to the east, with a cumulative throw of 65 m. The fault system continues eastwards until it disappears under younger Holocene lava flows. Some of the fissures and faults cut across the Holocene lavas, but reveal much smaller fault displacement, in the order of 1–12 m. This shows that the faults are not formed in a single event, but grow by repeated movement on the same fracture over a substantial period of time. The fissures and the faults are absent in the historical lavas nearby, which suggests a lull in the tectonic activity in this part of the fissure swarm over the past 1100 years. Although the system may lie dormant for now, it is still considered active.

Búrfell [64.0332, −21.8303] is a spatter-cone volcano situated near the middle of the Hjallar fault system about 1.8 km southeast of the point where the Heiðmörk road turns sharply to the northwest. The cone is well within walking distance from the road, an appealing and informative hike because the trail follows a drained lava channel that was full to the brim of red-glowing lava about 8100 years ago.

Búrfell is a near-circular cone with ramparts made up almost entirely of spatter ejected by lava fountains (Fig. 2.6). The crater rims rise up to 80 m above their surroundings and circumscribe a 140 m-wide and 60 m-deep

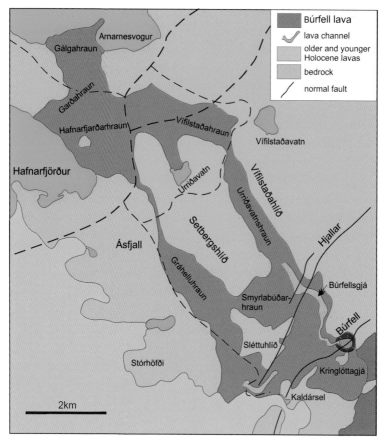

Figure 2.6 Búrfell spatter cone and lava flow, showing the key structures, distribution and local names for the lava. The flow covers a total area of 18 km2 and its volume is 0.35 km3.

crater, which contained a lava lake during the eruption. One of the northeast-trending Hjallar faults dissects the cone, and the eastern cone rims have dropped 2–5 m because of movements on this fault. The fault continues into the basement rocks to the north and south of the cone, where the vertical displacement across the fault is much greater (about 12 m). Thus, the Búrfell magma followed an existing fault to the surface, and that additional displacement occurred on this fault some time after the cone was built.

The lava that emanated from the Búrfell cone flowed to the northwest for about 12 km before entering the ocean in Hafnarfjörður and Skerjafjörður. The geological name of the flow is the Búrfell lava, but, in line with established Icelandic tradition, various sectors of the flow take names from nearby landmarks. Thus, locals may refer to the Búrfell lava by up to nine different names, much to the annoyance of visitors (Fig. 2.6). One sector, Gálgahraun ('gallows lava'), reminds us of a rather unpleasant part of Iceland's history. As the name implies, it hosted the gibbets used to execute serious offenders in the past.

The Búrfell lava consists of three main branches, and their relative positions tell us how the eruption progressed. The first to form was the Kaldársel branch. At the beginning of the eruption, the lava flowed in an open channel to the south, feeding a parasitic (secondary) lava pond that formed at present-day Kringlóttagjá (Fig. 2.6). By transport via internal lava pathways or tubes, the Kringlóttagjá pond fed inflating pāhoehoe lavas advancing down towards Kaldársel and then to the northwest through the valley between Sléttuhlíð and Stórhöfði, where they disappear under the younger Gráhelluhraun branch of the Búrfell lava. The Kaldársel branch disappears under younger lavas to the south. The mechanism of lava inflation was first recognized by Sir George Stuart Mackenzie, Sir Henry Holland and Dr Richard Bright, when they examined the Kaldársel branch in 1810, a discovery that went unnoticed for the subsequent 180 years.

As the eruption continued, the Búrfell cone grew in height and so did the level of the lava lake in the crater. As a result, the lava breached the cone rims in the southwest, cascading down the outer slope and spreading lava westwards from the cone and forming Smyrlabúðarhraun. There it split into two parallel branches. The southern one was formed as the lava flowed through the corridor between Sléttuhlíð and Setbergshlíð, forming Gráhelluhraun, and the narrow strip of lava that extends all the way to downtown Hafnarfjörður, where it disappears under the lavas from the northern branch. The northern one (the largest branch) followed a direct northwest passage between Vífilstaðarhlíð and Setbergshlíð, then spread westwards to the coast in Hafnarfjörður and Skerjafjörður. In the process the lava blocked a small valley and formed the small Lake Urriðakotsvatn. The 3.5 km-long Búrfellsgjá lava channel was formed during this stage of the eruption, feeding lava to the northern branch. The surface of these two lava branches is to a large extent covered by rubble. In some places these branches exhibit structures, such as lava tubes [64.0717, −21.8926] and classic pāhoehoe flow margins, indicative of lava inflation [64.0750, −21.9041] (e.g. Fig. 2.6: sites 1 and 2).

Locality 2.4 Tvíbollahraun and Kapelluhraun, the only historical lava flows in the greater Reykjavik area

On the outskirts of Hafnarfjörður, driving west on Highway 41, are the only historical lava flows within the urban limits of greater Reykjavík (see Fig. 2.1). The one closer to Hafnarfjörður is called Hellnahraun ('pavement lava' [64.0376, −21.9769]) and was formed by an eruption at the Tvíbollar cone row in about 950. The vents are out of sight, at a locality called Grindarskörð, and are within the Brennisteinsfjöll volcanic system. Thus, the lava travelled 17 km before coming to a halt here at Hvaleyrarholt, flowing across the Krýsuvík volcanic system in the process (see Fig. 1.5). It is pāhoehoe lava and it features exceptionally well-preserved surface structures, such as tumuli and lava-rise pits, indicative of lava inflation. The tumuli are easily recognizable, as mounds bounded by raised crustal slabs and dissected by a central cleft. The other one is *Kapelluhraun* ('Chapel lava') [64.0245, −21.9697], also known as Nýjahraun ('new lava'), and is derived from a cone row located at the foot-hill of Undirhlíðar, about 7.5 km directly south of the main road. Kapelluhraun is an 'ā'a lava flow and was formed by an eruption in the twelfth century during a major volcanotectonic episode on the Trölladyngja volcanic system. An excellent outcrop through the Kapelluhraun lava is present at a small road-ballast pit near *Bruni* [64.0298, −21.9706], which not only reveals the classical 'ā'a nature of the lava but demonstrates that endogenous growth also played a role in its emplacement. At this site the 'ā'a features tumuli and lava-rise sutures, structures that are indicative of lava inflation and hitherto only identified in pāhoehoe. The Ögmundarhraun lava near Grindavík was produced by the same eruption as Kapelluhraun (see also p. 73).

Reykjanes–Grindavík–Kleifarvatn
The architecture of a mid-ocean ridge

To begin this excursion from Reykjavík, simply follow the route to the international airport at Keflavík; head south through town and then west on Highway 41.

Despite its inhospitable appearance, the Reykjanes Peninsula is seventh heaven for geologists, especially tectonics and volcano enthusiasts. Here vegetation cover definitely does not impede exposure, and rocks outcrop at every footstep. It is incomparable in the sense that nowhere on Earth is the intricate architecture of a mid-ocean ridge system better exposed or as accessible for intimate inspection. Its inimitability does not end there;

because it is probably the best example available to us that mimics the structure of the very first landmasses formed on Earth, and thus can be viewed as a window into the deepest past of geological history.

The Reykjanes Volcanic Belt consists of four northeast-trending volcanic systems arranged in a step-wise fashion across the peninsula. The mountain range along the centre of the peninsula roughly delineates the axis of the belt. The steady westward drop in altitude from the Hengill volcanic system, at 803 m above sea level, to the Reykjanes system (243 m), represents decreasing influence of the mantle plume. The buoyancy effects of the plume gradually taper off with increasing distance from the centre of the mantle plume. The mountain range consists of subglacial and submarine volcanic structures, such as móberg ridges and table mountains, flanked, and in parts overlain, by subaerial lavas formed during the Holocene and to a lesser extent in historical times (see Fig. 2.1).

Disregarding the area from Vogarstapi to Garðskagi, the northwestern part of the Reykjanes Peninsula consists of three abutting Holocene lava shields, which are, in order from east to west, the Hrútagjá, Þráinskjöldur and Sandfellshæð. A similar arrangement is found on the southeast side of the peninsula, where six lava shields occupy similar positions (Geitahlíð, Herdísarvík, Selvogsheiði, Heiðin há, Leitin and Tröllahlíð).

Hrútagjá [64.0340, −22.0979] and *Þráinskjöldur* [63.9618, −22.3534] are abutting half shields with their summits at 200–240 m above sea level and tacked onto northwest slopes of the móberg mountain range. They produced pāhoehoe lavas that spread radially northwards and form the 20 km-long coastline between Straumsvík and Vogarstapi, some 10–11 km from the source vents. The Hrútagjá shield lavas formed 6–7000 years ago and constructed some of the largest tumuli found in Icelandic lavas and features excellent outcrops through such structures, as well as inflated pāhoehoe sheet lobes, at the *Rauðimelur quarry* [64.0316, −22.0653]. Þráinskjöldur was formed 14600 years ago and is a significantly bigger lava shield. It is heavily dissected by fissures and faults formed by rifting on the Reykjanes volcanic system. These faults are easily seen from the road in the form of northeast-trending linear scarps on the lower slopes of the shield. Farther up slope is the pyramid-shaped móberg cone Keilir, an island amid an ocean of lava. The third lava shield, *Sandfellshæð* [63.8602, −22.5747], formed 13600 years ago and has a more regular shape. It covers a large portion of the southwest corner of the peninsula. It is an exceptionally gently sloping shield (<

3°) that rises to 90 m above sea level at the summit and its lavas cover 120 km². Sandfellshæð has a well-formed summit crater, 450 m wide and 20 m deep, which is located about 4–5 km east of the road (Fig. 2.1).

All three shields are formed during the earlier part of the Holocene, between 6000 and 15000 years ago. Their cumulative volume is 15 km³, accounting for more than 75 per cent of the volume of magma erupted onto the surface in this sector of the peninsula during the Holocene. They also contributed significantly to raising the area out of the sea. The original distribution of the shield lavas is partly obscured by younger Holocene and historical lava flows; examples include the historic Hellnahraun and Kapelluhraun lavas (mentioned earlier), as well as the youthful-looking Afstapahraun lava flow, which entered the sea some 7.5 km farther to the west. This 14 km-long 'ā'a lava originated at a 3 km-long cone row near Trölladyngja, and it fills in the low where the flows of the Hrútagjá and Þráinsskjöldur lava shields meet.

The Quaternary basement that outcrops from *Vogarstapi* [63.9712, −22.4138] across *Miðnes* [63.9970, −22.6462] to *Garðskagi* [64.0810, −22.6895] is the oldest rock formation to the west of Reykjavík. It is mostly composed of glacially eroded lavas formed by one lava shield volcano or more. The easternmost part (i.e. Vogarstapi) is dissected by young fissures and faults formed in association with activity on the Reykjanes volcanic system. Directly south of Vogarstapi are the remains of *Stapafell* [63.9055, −22.5247] and *Þórðarfell* [63.8909, −22.5214], two submarine volcanoes composed of strongly olivine phyric pillow lava and associated volcani-clastic deposits formed at the end of the Weichselian glaciation. These volcanoes have almost been removed by excavation. *Rauðimelur* (the second one with that name) is the name of a 3 km-long spit bar [63.9286, −22.4808] that extends to the northwest from Stapafell. The bar was formed by marine erosion of two volcanoes, as is evident from its composition – basalt clasts mixed in with olivine crystals.

Locality 2.5   Mid-ocean ridge rising out of the sea

If you travel west on Highway 41 towards the international airport at Keflavík, as you pass the town of Njarðvík follow the signs to the town of *Hafnir* [63.9339, −22.6838] (Road 44). Head directly south on Road 425, through Hafnir, towards the geothermal plant at Reykjanes. From there follow the signs to *Reykjanesviti* [63.8155, −22.7043] (the lighthouse at

Reykjanes) and, passing the lighthouse, you have reached the destination (see Fig. 2.1).

Reykjanes [63.8128, −22.7147] is the southwestern outpost of Iceland and the point where the Mid-Atlantic Ridge rises out of the sea. On land the ridge crest (i.e. the rift valley) is defined as a distinct fault zone delineating a shallow 10 km-wide composite graben structure trending northeast from Reykjanes to Vatnsleysuströnd. Along the centre of the rift is a series of 40–100 m-high steep móberg hills composed of pillow lava and hydroclastic tephra and breccia, including *Bæjarfell* [63.8155, −22.7043], where the present lighthouse stands, and *Valahnúkur* [63.8108, −22.7114], where the foundations of the old lighthouse can still be seen. On both sides, lava shields border the móberg hills, which in turn are partly covered by younger Holocene and historical lava flows (see Fig. 2.1). But not all of Reykjanes is volcanic. The cove south of Valahnúkur features a spectacular boulder beach formed by the rampant storms of the North Atlantic, and farther to the north the black sand dunes at Stóra-Sandvík are a blunt reminder that the wind blows fearlessly through Reykjanes. It also hosts a small (~1 km^2) high-temperature geothermal area with small steaming vents and bubbling solfataras. However, these hot springs are different from what they used to be, partly because the area is now utilized for power generation.

The móberg mountains of *Sýrfell* [63.8369, −22.6598], Bæjarfell and Valahnúkur were formed by submarine fissure eruptions when sea level was standing about 70 m higher than today, during the waning stages of the Weichselian glaciation. These móberg mountains consist of pillow lava, breccias and tuffs, which in the case of Sýrfell are capped by small scoria cones and thin lava flows, indicating that the volcano rose out of the sea to form a small island. However, Valahnúkur volcano seems to have remained fully submerged during its period of activity, as it features a basal sequence of pillows capped by a breccia unit and another pillow-lava sequence. The hill itself is split in the middle by a 20 m-wide graben structure, and the walls on either side provide an excellent outcrop of the pillows and the breccia units (Fig. 2.7a). The youngest tephra layers in the **regolith** that laps onto the eastern slopes of Valahnúkur are from a volcanotectonic episode that raged at Reykjanes in the thirteenth century, suggesting that this graben structure is very young and was, at least partly, formed in historical times.

The youngest volcanic formation at Reykjanes is the 4.5 km-long *Yngri Stampar* [63.8197, −22.7220] cone row, along with the lavas and the tuff it produced, which now cover the northern part of the point (see Fig. 2.1). These were formed in the so-called Reykjanes Fires, a major volcanotec-tonic episode within the Reykjanes system between 1210 and 1240.

Vatnsfell [63.8150, –22.7234] is a rather inconspicuous rise on the coast north of Valahnúkur, but is worth a visit for those interested in having a closer look at the internal make-up of tuff cones (Fig. 2.7). It is made up of lapilli tuff deposits, which are the remnants of two tuff cones formed by submarine explosive eruptions in the early stages of the Reykjanes Fires in about 1210. The internal structures of these cones are best exposed in out-crops opposite Karl, the sea stack rising about 300 m offshore. The older cone, which makes up the bulk of Vatnsfell proper, was 30 m high and 650 m wide, with the crater located 100 m offshore. The lower part of the older tuff cone consists of wavy to cross-bedded ash deposits formed by repeated base surges, whereas the upper part consists of coarser-grained lapilli tuff representing base surge and tephra fall units. On either side of Vatnsfell, the older cone is capped and flanked by the deposits from the younger cone, which consists of fine-grained lapilli tuff with repeated pairs of base-surge and tephra-fall layers (Fig. 2.7b). Many impact craters formed by ballistic blocks ejected from the craters during the eruptions commonly disrupt bedding in both cones. The younger tuff cone was a much larger structure than its earlier counterpart, at least 55 m high and with a basal diameter of about 1600 m. The main crater was located about 400 m off-shore near the current position of sea stack Karl. This cone is probably the mountain, described in the thirteenth-century chronicles, that rose out of the sea of Reykjanes in 1211. The lavas of Yngri Stampar were formed either in the same eruption or the second eruption of the Reykjanes Fires in 1223. Some of the feeder dykes dissect the Vatnsfell tuff connecting to the overlying flows of the Yngri Stampar cone row (Fig. 2.7c).

The thirteenth-century chronicles also mention an explosive eruption of the shore of Reykjanes in 1226, which caused widespread tephra to fall on land. This tephra layer is known as the Middle Ages Tephra (Miðaldarlagið) and is found in soils all over the Reykjanes Peninsula, including Reykjavík and the Esja region. It forms an important time marker and has been used extensively for dating historical lava flows in the area. Eldey, a steep-sided islet 14 km offshore, is yet another remnant of an emergent submarine volcano formed by the Reykjanes Fires, and it may be the source vent for the Middle Ages Tephra. Now it hosts the largest gannet-breeding ground in

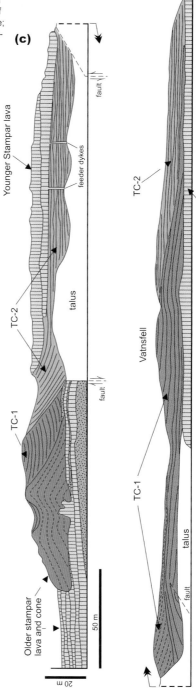

Figure 2.7 (a) (opposite) Pillow lava at Valahnúkur; **(b)** (opposite) lapilli tuff beds in the younger Vatnsfell tuff cone; **(c)** cross section showing the arrangement of the Vatnsfell tuff cones and Stampar feeder dykes at the coast of Reykjanes.

(c)

Younger Stampar lava

feeder dykes

fault

fault

talus

TC-2

TC-1

Older stampar lava and cone

50 m

20 m

TC-2

Skálafell lava

Vatnsfell

TC-1

talus

fault

the Northern Hemisphere and its top is completely covered by a knee-deep layer of guano.

The chronicles also mention eruptions at Reykjanes in 1231, 1238 and 1240. The last eruption made the Sun appear 'red as blood', which implies volcanic plumes rich in sulphuric aerosols, and atmospheric perturbations on a regional scale. Such plumes can be generated by effusive basalt eruptions, provided that the volume of erupted magma is reasonably large. The only historical eruptions that fit the requirements of correct setting, age and size are those that produced the Arnarsetur and Illahraun lavas north of Svartshengi, some 18 km to the northeast of Reykjanes (see Fig. 2.1). Tephrochronology shows that these two lava flows were formed several years after the Middle Ages Tephra fell in 1226 and thus may represent the final episode in the Reykjanes Fires.

The second-youngest volcanic formations at Reykjanes are the *Eldri Stampar* [63.8250, −22.7172] and *Tjaldastaðargjá* [63.8528, −22.6553] cone rows and lavas formed by another volcanotectonic episode 1500–2000 years ago. Other fissure volcanoes in the area are of unknown age but pre-date the volcanoes mentioned above, illustrating that fires have raged at Reykjanes periodically throughout the Holocene. The oldest Holocene formations at Reykjanes are the lava shields Sandfellshæð, *Skálafell* [63.8125, −22.6822] and *Háleyjarbunga*, [63.8166, −22.6510] which together are the largest lava formations in the area and solely composed of pāhoehoe flows. Háleyjarbunga may hold a special interest for petrology enthusiasts, because it is composed of picrite basalt and is exceptionally rich in olivine phenocrysts. Some of the lava outcropping in the walls of the summit crater and the cliffs along the southern coast is green with olivine.

On the way
Continue on Road 425 along the southern coast towards the town of *Grindavík* [63.8453, −22.4351]. The route will take you through several older Holocene lava flows and, as you approach the putting greens of the Grindavík golf course, the mountain Þorbjarnarfell [63.8641, −22.4405] will appear on the skyline to your left. It is the highest (243 m) subglacial móberg volcano within the Reykjanes volcanic system and is noteworthy for the spectacular graben structure that cuts across its top. On the other side of Þorbjarnarfell is *Svartshengi* [63.8801, −22.4319], the geothermal field that supplies the communities on the Reykjanes Peninsula with hot

water for domestic use and house heating, but probably better known for its spa, the Blue Lagoon. It is an ideal spot for a relaxing dip in the warm mineral-rich water to recharge the batteries for the second half of the excursion.

Return to Road 245 travelling east towards *Krýsuvík* [63.8869, –22.0651], which initially traverses the lavas of Sundhnúkur and Vatnsheiði. The former lava flowed into the sea to form the point Þórkötlunes and the inlet that hosts the harbour at Grindavík. It is one of the larger Holocene fissure-fed flows on Reykjanes (volume 0.6 km³) and it poured out of an 8.5 km-long fissure about 2400 years ago. The olivine-rich Vatnsheiði lavas are pāhoehoe flows derived from three small picrite lava shields northeast of Grindavík that were constructed by eruptions during the early Holocene. The Vatnsheiði lavas host abundant gabbro xenoliths and large (2–3 cm) feldspar-dominated xenocrysts, which are readily discernible in the coastal outcrop at *Hrólfsvík* [63.8494, –22.3661] (Fig. 2.1: site 9).

A little farther to the east, where the road climbs a steep incline, is *Festarfjall* [63.8573, –22.3373], at 100 m high the tallest seacliff on the Reykjanes Peninsula, which consists of weakly consolidated hydroclastic tephra and is capped by a lava flow. The feeder conduit that fed the summit lava can be seen as a dyke dissecting the tuff sequence in the cliff facing the sea. The mountain's name is derived from the dyke, which was thought by locals to be the chain anchoring the mountain to the bedrock; hence, the name Festarfjall ('anchored mountain'). In geological terms, Festarfjall is a remnant of a submarine table mountain, a Surtseyan volcano that was formed in an eruption towards the very end of the Weichselian glaciation. A word of warning: rockfall is a common occurrence at Festarfjall and therefore we do not recommend hiking along the sandy beach in front of the cliff face. Northeast of Festarfjall is the 350 m-high table mountain, *Fagradalsfjall* [63.8961, –22.2934], with a cap of subaerial lava indicating that the Weichselian glacier was at least 300 m thick in the area when the volcano was formed.

Locality 2.6   The Ögmundur lava

In passing the southern end of the móberg ridge Vesturháls, one enters the rugged lava flowfield of Ögmundarhraun, renowned for the ruins of the old Krýsuvík homestead laid to waste by the lava in the twelfth century (Fig. 2.8). After the eruption, the homestead was rebuilt farther to the east at

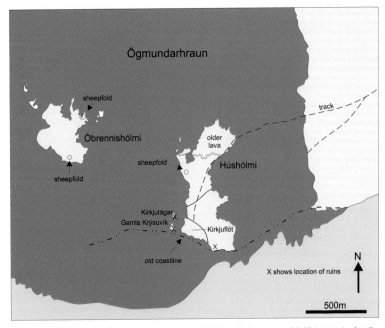

Figure 2.8 Map showing the ruins of the old Krýsuvík homestead laid to waste by the Ögmundarhraun lava (purple) in the twelfth century.

Bæjarfell. It is also one of the few lava flows in Iceland that takes it name from a person. The folklore tells the story of Ögmundur 'the Berserk', who cleared a passageway across the lava for the squire at the Krýsuvík estate, and upon completion of the 'road construction' he was to be rewarded by marriage to the squire's daughter. However, the squire had no intention of giving his daughter away to the peasant and had Ögmundur killed just as he was about to complete the job. Since then the lava has been known as Ögmundarhraun.

Ögmundarhraun is one of several lava flows formed in the volcano-tectonic episode, known as the Krýsuvík Fires, which raged within the Trölladyngja volcanic system between 1151 and 1188. The lava flows, including *Ögmundarhraun* [63.8359, −22.1655] in the south and Kapell-uhraun in the north, cover 36 km² and were produced by a series of volcanic fissures trending northeast for 28 km across the peninsula (see Fig. 2.1).

The old homestead of Krýsuvík was apparently a prosperous estate in its time, with a church and also rich fishing grounds offshore. The ruins, including the homestead, the church, and two sheep pens, are clearly visible in the kipukas Húshólmi and Óbrennishólmi, and can be accessed by walking down along the eastern margin of Ögmundarhraun and then following the path across the lava flow to Húshólmi.

Locality 2.7   The Krýsuvík area

Continue onwards to the east and then turn north on Road 42 towards Kleifarvatn. About 1 km up the road is the *Grænavatn* [63.8848, −22.0536], the largest explosion crater of a small maar volcano complex that features eight craters in total. The Grænavatn maar is about 300 m in diameter and 44 m deep. The stratigraphy of the inner wall of the maar suggests that a fountain-fed lava from an adjacent spatter cone flowed over an active geothermal field and sealed it. Consequently, the geothermal system became pressurized and eventually burst through the lava to form the Grænavatn maar. The Grænavatn maar is well known for olivine-gabbro xenoliths, and their occurrence in the fountain-fed lava suggests that they were carried to the surface by the magma erupting via the spatter cone.

Lake *Kleifarvatn* [63.9263, −21.9744] (9.7 km^2) is one of the deepest lakes in Iceland (107 m) reaching below sea level. It sits in a fault-bounded basin between móberg ridges Sveifluháls and Vatnshlíð and, like many other lakes in Iceland, it has no surface runoff. The móberg ridges on either side of Kleifarvatn were built up by a series of effusive and explosive subglacial eruptions, as is evident from the pillow lavas and tuffs that appear in outcrops along the road. At the foothills of Sveifluháls are the hot springs and the old sulphur mine at Krýsuvík. An earthquake took place beneath Kleifarvatn in the year 2000, opening a crack in the lake bottom. This crack drained so much water from the lake that it temporarily reduced its surface area by over 20 per cent. The lake has since recovered and the crack now features vigorous geothermal activity at the southern end of the lake. Return to Reykjavík by continuing north on Road 42, which intersects Highway 41 at the southern border of Hafnarfjörður.

View of the beginning phase of the 1973 Eldfell eruption on 24 January 1973 (upper panel) and the Syrtlingur phase of the 1963-67 Surtsey eruption in Vestmannaeyjar archipelago on 6 June 1965 (lower panel). Photographs by Sigurjón Einarsson.

Chapter 3

The south

General overview

South Iceland is bounded on either side by the active West and East Volcanic Zones (Fig. 3.1a). The South Iceland Seismic Zone transects the lowlands from the Hekla volcanic system in the east to the Brennisteinsfjöll volcanic system in the west (see Fig. 1.2). It is the source of some of the largest earthquakes in Iceland, with events between 6 and 8 on the Richter scale occurring periodically every hundred years or so (Fig. 3.1b). The most recent major events occurred in the years 2000 and 2008. At the surface, the South Iceland Seismic Zone is characterized by north–south trending strike-slip faults, but is thought to represent a major east–west trending transform fault linking the two volcanic zones.

South Iceland is characterized by contrasting topography, where the foreground is the vast flat-lying southern lowlands that are surrounded on the periphery by rugged mountains rising sharply up to 1000–1500 m. During glacial times the weight of the ice pushed the land down, so that in places it was well below sea level. When the Ice Age (Weichselian) glacier retreated from South Iceland, starting about 15 000 years ago, the lowlands were temporarily inundated by the sea and became the bottom of the then vast South Iceland Bay (Fig. 3.2). During the early Holocene the land gradually emerged from the sea free of the weight of glacial ice, and vast sandur plains developed as rivers emerging from the interior highlands dumped their sediment load onto the flat coastal plains. The Great Þjórsá lava flow that now covers the area between the Þjórsá and Hvítá-Ölfusá River made a significant contribution to this process some 8500 years ago (Fig. 3.1a).

The Great *Þjórsá lava* [63.9508, −20.6928] is the oldest flow of the Tungnaár lavas, a series of eight lava flows that emanated from fissures within the Veiðivötn volcanic system to the north of Landmannalaugar

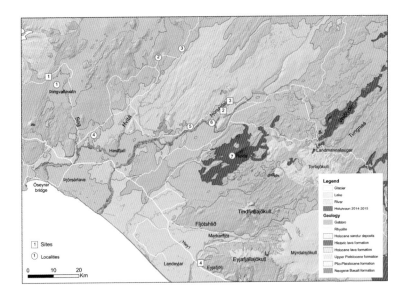

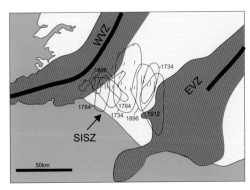

Figure 3.1 South Iceland. **(a)** The main geological features and excursion routes, stops and sites mentioned in the text; **(b)** occurrence and extent of damage by earthquakes on the South Iceland Seismic Zone.

during the Holocene and flowed to the southwest across the highlands and onto the Southern Lowlands. It is the largest of a series voluminous fissure-fed basalt-lava flows that were erupted in Iceland during the earlier part of the Holocene, with an estimated volume of about 25 km³. The largest lava-shield volcanoes from the same time period are of similar size (e.g. *Skjaldbreiður* [64.4088, −20.7521], 13.5 km³). The presence of these large-volume lava flows in the earlier part of the Holocene indicates that production rate of basaltic magma was somewhat greater then (by factor > 2) than at present. This higher production rate has been linked to the rapid

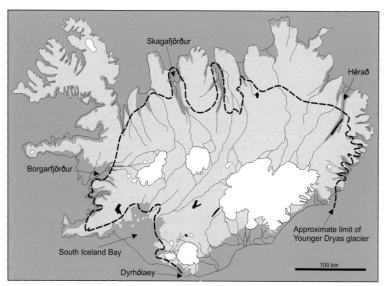

Figure 3.2 Distribution of lowland areas invaded by the sea towards the very end of the Weichselian glaciation. The extent of the Younger Dryas glacier is indicated by the dashed line.

crustal rebound after removal of the glacier, up to 1000 m thick, of the most recent glaciation, although this notion has yet to be tested.

The mountains on the periphery of the region are either currently active central volcanoes or volcanic landforms created by subglacial eruptions during the Pleistocene. The view from *Óseyrarbrú* [63.8786, −21.2123] ('Óseyri Bridge') is one of the most breathtaking sights in Iceland, revealing a chain of towering central volcanoes from south to north: *Eyjafjallajökull* [63.6276, −19.6359], *Mýrdalsjökull* [63.6365, −19.1164], *Tindfjöll* [63.7953, −19.5907], *Torfajökull* [63.9487, −19.1399] and *Hekla* [63.9951, −19.6504]. In the background to the north is a glimpse of *Langjökull* [64.6718, −20.0869] and *Hofsjökull* [64.8134, −18.8063], the second and third largest glaciers in Iceland, the latter also being the largest active volcano in the country.

We present two excursions in South Iceland. The first is sometimes nicknamed 'the golden triangle' because it is centred on *Þingvellir* [64.2561, −21.1284], the original site of the Icelandic Alþing ('parliament'), the *Great Geysir* [64.3104, −20.3031] – the grandfather of all erupting hot springs – and *Gullfoss* [64.3255, −20.1269] ('golden waterfall'), Iceland's most celebrated waterfall. The second excursion is to the Hreppar region, including Hekla, the queen of Icelandic volcanoes, and *Þjórsárdalur* [64.1216, −19.8712], the region laid to waste by the first historical eruption at Hekla in 1104.

Þingvellir–Geysir–Gullfoss

On the route from Reykjavík to Þingvellir, the first part of the trip takes you through the older and deeply eroded Pliocene to Pleistocene succession, outcropping in the hills Úlfarsfell and Helgafell and, farther away, in Mt Esja (see Chapter 2). Along the road is the 17 km-long hot-water pipeline running from the geothermal area at Reykir in Mosfellsbær to Reykjavík, which has provided the city dwellers with thermal water for space heating since 1933.

As the road turns inland at the bottom of Leiruvogur and climbs up onto the Mosfellsheiði moor, the gently sloping *Borgarhólar* [64.1519, −21.4673] lava shield, one of the source volcanoes for the Grey Basalt foundation in Reykjavík, comes into view towards the southeast. Another lava shield, *Hæðir* [64.1731, −21.2702], is on the eastern horizon. At the foothills of Mt Esja to the north of the road, the large gabbro body outcrops, representing the roots of the *Stardalur* [64.2203, −21.5411] central volcano. The volcano also produced the Móskarðshnjúkar rhyolite that caps the Mt Esja sequence in this area, and shines brightly if there is a break in the cloud cover.

Locality 3.1
Þingvellir: the heart of Icelandic culture and heritage

No trip to Iceland is complete without a visit to the national park at Þingvellir. This historic site offers not only superb geology but also a glimpse into Iceland's history. Þingvellir is the original site of the Alþing, the general assembly of the Icelandic commonwealth (930–1262). It was established in 930 and is still in operation today, making it Europe's oldest operating parliament. In the days of the commonwealth, each year the chieftains and their followers gathered at Þingvellir to settle their debates. The laws of the land were recited by the law speaker from Lögberg – a natural pedestal made of lava – using the walls of Almannagjá to echo his words over the constituency.

Þingvellir is truly the heart of Icelandic culture and it has been the venue for some of the most significant events in the country's history, perhaps none as significant as the nationwide adoption of Christianity in the year

1000, which came about through political debate at the Alþing. The main reason appears to be that a majority wanted to maintain peace and unity throughout the nation and to prevent civil war between the extremists within opposing factions, the followers of the old heathen religion and Christians. The records of the event also offer a rare indication of the contemporary understanding of volcanic eruptions.

As the members of the Alþing were debating the adoption of Christianity, news was brought of an eruption in the Ölfus district, South Iceland. It was apparent that the lava flow would overrun the farm of the heathen priest Þóroddur, and the firm among the heathen followers spoke: 'We are not amazed that the heathen gods are enraged at such a decision.' Then the heathen chieftain Snorri replied: 'At what were the gods enraged when the lava on which we are now standing was formed?' According to the chronicles, this reply was the turning point in the debate and ended the protest by the heathen extremists. It also shows that volcanic eruptions were generally viewed as phenomena of nature rather than punishment from higher authorities. For further information on the other historical aspects of Þingvellir, visit the information centre at the park entrance.

Viewing the Þingvellir area from the main lookout (see Fig. 3.1: site 1), it is immediately clear that the geology of the area is one of seafloor spreading, displaying the intricate association of crustal rifting and volcanics within the West Volcanic Zone. The bedrock in the area consists of Holocene basalt lavas covering the central part of the fault-bounded lake basin, known as the Þingvellir graben, and Pleistocene pillow lavas and **hyaloclastites**, forming the móberg ridges and table mountains aligned along the periphery of the basin. The Þingvellir graben is completely circumscribed on all sides by volcanoes that belong to four active volcanic systems, the Prestahnúkur and Hrafnabjörg systems to the north and the Hengill and Hrómundartindur systems to the south (see Fig. 1.5).

Starting in the north, the first in view is the strongly faulted *Ármannsfell* [64.3189, −21.0338] mountain, which was formed by a subglacial eruption that broke through the Weichselian glacier to form a table mountain composed of strongly olivine phyric basalt. Behind Ármannsfell to the right is Skjaldbreiður, the prototype of monogenetic lava shields. Exhibiting a perfect shield shape, the volcano has a basal diameter of 16–18 km and stretches right across the northern end of the Þingvellir graben (see Fig. 1.15a). It is characterized by gently rising (1–8°) outer slopes climbing from 350 m at the base to 1000 m at the summit. The Skjaldbreiður volcano was constructed by a long-lived effusive eruption that surely lasted for decades, and possibly several hundred years, and its lavas have a total volume of 13.5 km^3.

To the east on the edge of the graben are the subglacial volcanoes *Hrafnabjörg* [64.2723, −20.9272] and *Tindaskagi* [64.3353, −20.8165], displaying the classic forms of a table mountain and a móberg ridge, respectively. Farther east in the background are the mountains of the Laugarvatn region, where J. G. Jones (1969, 1970) established the characteristic facies associations for subglacial volcanoes in his classic studies during the late 1960s.

The extensive pāhoehoe lava flowfields that fill in the graben in front of Skjaldbreiður and Ármannsfell were mainly formed by effusive eruptions within the Hrafnabjörg volcanic system in the early Holocene. One is the eruption that constructed the Skjaldbreiður volcano, whereas the other three originated from relatively short fissures to the east of Hrafnabjörg mountain. Three of these lavas were formed during the early Holocene, issuing about 26 km^3 of lava into the Þingvellir graben.

The first to form was the ~10 200 year old Þingvellir lava (^{14}C age 9190 ± 65 years) that now floors the northern half of Lake Þingvallavatn and the area immediately to the north and east of the lake, and thought to originate from volcanic fissures to the southeast of Hrafnabjörg. The flows of this lava are exposed in the fault scarp at *Almannagjá* [64.2563, −21.1277], where they exhibit the classic structure of inflated pāhoehoe lobes. Model calculations show that it took more than a year to accumulate the vertical thickness of the succession exposed at Almannagjá. Every level in the succession consists of a series of laterally arranged sheet-like lava lobes, each formed by successive breakouts from an earlier lobe, so it is obvious that the Þingvellir lava was formed by a long-lived eruption that may have lasted for decades, if not centuries.

This activity was closely followed by the ~9 000–10 000 year old Skjaldbreiður lava-shield eruption, which, in addition to constructing the lava shield in the northern part of the graben, sent lavas southwards, which partly covered the existing Þingvellir lava. This episode culminated with the *Eldborgir* [64.2388, −20.9371] fissure eruption about 7 000–8 000 years ago, which produced the rubbly-surface flows that cover the eastern part of the Þingvellir lava. These eruptions had a major impact on the evolution of Lake Þingvallavatn, because they both reduced its size and changed its hydrological character (Fig. 3.3). The youngest lava, *Þjófahraun*

[64.3048, −20.8409] (^{14}C age 3360 ± 35 years), was formed by a fissure eruption 3000–4000 years later.

To the east, the Þingvellir graben is bordered by the interglacial (Eemian) Lyngdalsheiði lava shield and closed off to the south by the jagged mountains that make up the *Hrómundartindur* [64.0762, −21.2015] and *Hengill* [64.0772, −21.3131] central volcanoes (see Fig. 1.5), which were mainly constructed by repeated subglacial centralized eruptions during the Weichselian glaciation (Table 1.1). The irregular form of these central volcanoes is a good example of how the common conical shape can be modified by subglacial volcanism. The Hrómundartindur volcano has produced the most basic and some of the more evolved rocks in the Þingvellir area, namely the picritic pillow basalts of *Miðfell* [64.1782, −21.0518] and *Dagmálafell* [64.1766, −21.0546] (the name indicates that the sun stands above the mountain when morning passes and day begins, i.e. 9 am), as well as the andesites of *Stapafell* [64.0921, −21.1723]. Both central volcanoes are very much alive, as is clearly indicated by occurrence of Holocene eruptions and the many hot springs and fumaroles venting the high-temperature geothermal systems at *Nesjavellir* [64.1083, −21.2569] and *Ölkelduháls* [64.0593, −21.2337].

The *Tjarnahnúkur* [64.0650, −21.2158] scoria cone, a parasitic vent on the Hrómundartindur volcano, erupted in the early Holocene (10 000-10 500 years ago) to form a small basaltic pāhoehoe lava flow field, which reached the shores of a proglacial lake that marks the embryonic stage of Lake Þingvallavatn. During the Holocene, four small fissure eruptions have occurred within the Hengill volcanic system, with vent sites on the north flanks of the volcano and within its fissure swarm. The youngest and most prominent was the 1800-year-old eruption that formed the *Nesjahraun* [64.1393, −21.2137] lava and the double-crater tuff cone of *Sandey* [64.1837, −21.1654], the island in the middle of the lake. The Sandey cone represents the hydromagmatic phase of the Nesjahraun eruption, the only eruption known to have occurred within the boundaries of Lake Þingvallavatn. The tephra produced by the activity at Sandey is present in the soil cover to the east of the lake.

The area's most prominent feature is the spectacular northeast-trending Þingvellir graben nested within the northern limb of the Hengill fissure swarm. The graben, 10–20 km wide, narrows to the southwest, and is bounded by a series of normal faults on either side, the main ones being

Almannagjá in the west and Hrafnagjá to the east. The graben floor slopes to the southwest, reaching 600 m altitude around the Skjaldbreiður lava shield and dropping below sea level at Lake Þingvallavatn.

Perhaps the most dramatic structure at Þingvellir is the giant crack of Almannagjá, where the ground has simply been ripped apart by the plate movements. Almannagjá is 7.7 km long, its greatest width 64 m and its maximum throw 30–40 m. The opposing boundary fault, *Hrafnagjá* [64.2573, −21.1108], is 11 km long, 68 m where it is widest, and has a maximum throw of 30 m. These and other boundary faults of the Þingvellir graben are thought to be the surface expression of deep-rooted normal faults that extend through the crust, formed as the result of seafloor spreading. Within the Þingvellir graben are open fissures formed by the same process.

During the past 10 500 years the relative total vertical movement on the graben faults between Almannagjá and Hrafnagjá is in excess of 40 m (averaging about 1 mm/year), and the estimated horizontal extension is of the order of 70 m (averaging about 7 mm/year). Rifting within the graben is episodic, and individual tectonic events are associated with vertical and horizontal movements of the order of several metres. The most recent rifting episode occurred in 1789, when part of the graben floor subsided 1–3 m, with the consequence that the old parliament flats became unusable as a venue for the annual assembly of the Alþing. Since then the Alþing has resided in Reykjavík.

Þingvallavatn, the largest lake in Iceland, covers 82 km^2 and fills the deepest part of the graben. With a maximum depth of 114 m, it descends to 10 m below sea level. The present-day Þingvallavatn began to form around 11 500 years ago as a proglacial lake in front of the retreating Younger Dryas glacier (Fig. 3.3). As the glacier retreated farther, rivers emanating from glaciers in the Langjökull area discharged into the basin and consequently the lake grew to a size approximately equalling its present extent. The eruptions of the 10 200–7000 year old Þingvellir–Skjaldbreiður–Eldborgir lavas had a profound impact on the hydrological character of the Þingvellir graben because they effectively terminated the surface flow of glacial rivers from the north and reduced the lake size by more than 50 per cent (Fig. 3.3). Ever since, the glacial meltwater discharging from the southern end of Langjökull has percolated through the porous lava formations, undergoing a natural filtering in the process, before discharging as freshwater springs into the lake. As a result of this steady groundwater-driven discharge and

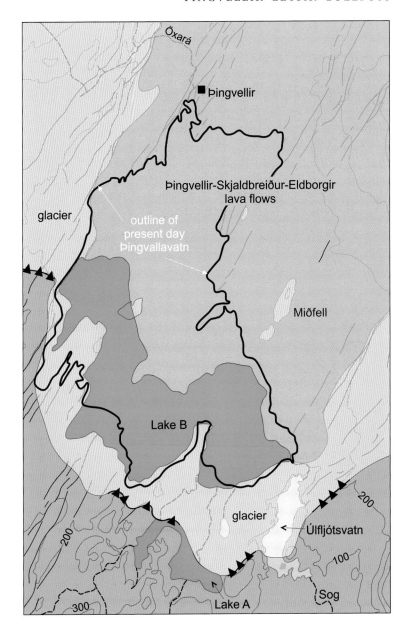

Figure 3.3 Evolution of Lake Þingvallavatn during the Holocene: lake A is the proglacial lake stage formed as the late Weichselian glacier retreated from its maximum stand in early Pre-boreal times; lake B inferred extent of the lake at about 10200 years ago when the northern part of the Þingvellir graben was flooded by the voluminous Þingvellir–Skjaldbreiður–Eldborgir lava flows, changing it to a spring-fed lake. Since then the lake has gradually enlarged because of consistent subsidence within the graben.

continuing subsidence of the graben floor during the Holocene, the lake has gradually grown to reach its current size.

To reach the next locality, follow Road 36, which takes you across Hrafnagjá, the fault that marks the eastern margins of the Þingvellir graben. Then turn at *Gjábakki* [64.1969, −21.0275] and follow Road 365 over the *Lyngdalsheiði* [64.1536, −20.8886] interglacial lava shield. The route across Lyngdalsheiði provides good views of the internal architecture of the sub-glacial volcanoes of the Laugarvatn Formation, with pillow lava and the capping-tuff sequence of the *Kálfstindar* [64.2520, −20.8576] móberg ridge appearing first, followed by the lava-crowned table mountain of *Laugarvatnsfjall* [64.2295, −20.7737]. At *Laugarvatn* [64.2177, −20.7320], geothermal springs occur along the northern shore of the lake and are utilized for heating natural saunas, which provide the ultimate perspiration experience. From Laugarvatn follow Road 37/35 eastwards to the Geysir geothermal field.

Locality 3.2   Geysir geothermal field

Geysir [64.3104, −20.3031] is perhaps the world's most famous spouting hot spring and its reputation is such that in many languages such springs are known as geysers. Geysir is situated among other hot springs in a field of silica sinter deposit at the eastern end of the hill Laugarfjall. Most of the springs here are alkaline with clear water. Geysir is by far the largest spring and it consists of a 3 m-wide and 22 m-deep shaft that opens into a 2 m-deep and 15 m-wide bowl-shaped basin at the summit of the regular silica sinter dome that surrounds the hot spring. The hot springs in the Geysir area are first mentioned after the South Iceland earthquake in 1294. It is possible that the Great Geysir was rejuvenated by this event, but the 3100-year-old Hekla tephra layer that sits immediately beneath the silica sinter dome indicates a considerably longer lifetime for Geysir.

Geysir eruptions are initiated by superheating of the water column at about 10 m depth in the shaft, and the onset of an eruption is usually indicated by rumbling in the ground, vigorous boiling and periodic inflation of the water in the bowl, forming a large cupola-shape bubble. As the bubble bursts, the eruption begins by sending up a series of water jets, growing in height until the event has reached its maximum. In its prime, Geysir produced 30–80 m-high jets in eruptions that lasted 5–10 minutes. In 1750 the Icelandic naturalists, Eggert Ólafsson and Bjarni Pálsson, described a Geysir eruption as follows:

We stayed by Geysir in case it would erupt. Earlier we had thrown several boulders of silica sinter, which lay astray, down into the bowl of Geysir. At first we heard a heavy thud beneath us. It resembled the sound of cannon fire from a distance. In all, the thuds were five and of increasing intensity. At the same time the ground jolted, like it was going to be lifted or ripped open. Following the sixth thud, the first jet emerged from Geysir. Thereafter the intensity of the eruption increased with each thud and the water was ejected as columns. The boulders we had thrown into the bowl had split into smaller fragments. The largest of these fragments were close to fist size. These fragments were ejected with the water and some reached heights greater than their associated water columns. We located ourselves up wind from Geysir so we could best observe the eruption because the steam blocked the view on the downwind side. This was also done to avoid injuries and burns from the hot water falling down from the columns. From the beginning of this eruption we noticed that, not only was the water lifted into the air by each emerging jet, but also simultaneously the total volume of water retained within the crater was inflated and overflowed the banks of the bowl, mostly to the north. . . . This eruption was equivalent to the highest and greatest eruptions known at Geysir. Although it did not reach the height of Laugafell, a small hill, 70 fathoms [116 m] high, in the neighbourhood of Geysir. In this particular eruption the water columns rose to approximately 60 fathoms [100 m]. According to the inhabitants of Haukadalur, the height of the water column in some Geysir eruptions reaches heights equal to that of Laugafell, but eruptions of such intensity occur only before the arrival of heavy rainstorms. The eruption lasted for ten minutes, with jets ejected every three seconds, and the eruption featured around 200 jets in total.

There are many other hot springs in the Geysir area, such as Strokkur ('the churn'), which has been known to have erupted jets up to 100 m high (see Fig. 1.18), Smiður ('the smith'), Óþerrishola ('the drizzly hole'), and Sóði ('the slob'). The most beautiful spring in the area is without doubt Blesi, which consists of two holes, one with bluish and the other greenish crystal-clear water; this effect is caused by colloids in the water.

On the way

Heading east from the Geysir area on Road 35 takes you across cultivated flat-lying farmlands and then up along the western edge of the Hvítá gorge and straight to the *Gullfoss* [64.3255, −20.1269] waterfall.

Locality 3.3   Gullfoss and Hvítárgljúfur

Where the glacial River Hvítá descends from the highlands onto the Southern Lowlands, it has eroded an impressive gorge into the edge of the highlands, along faults trending north-northeast. The Hvítá gorge (Hvítárgljúfur) is 2500 m long and its maximum depth is 70 m. The two-step waterfall of Gullfoss is at the head of the gorge and has a total drop of 32 m. Although the course of the river is controlled by faults, Gullfoss and the gorge owe their existence to the contrasting lithologies that make up the succession in the Gullfoss area. The succession is characterized by fairly thick sedimentary units, each capped by somewhat thinner lava flows (Fig. 3.4). The lavas are resistant to erosion and represent the formations that carry the waterfall. The sedimentary units are softer than the lavas and are therefore eroded faster by the water. This causes under-cutting of the lavas, forming an overhang, which eventually collapses, and the process is repeated. This step-wise process is thought to have carved the gorge into the landscape in less than 10 700 years, at an average rate of 25 cm/year. An alternative view is that the gorge was to a large extent formed in catastrophic floods (i.e. jökulhlaups) that were produced in association with the retreat of the Younger Dryas glacier in Pre-Boreal times (i.e. 11 700–10 500 years ago).

From Gullfoss follow Road 35 back to Geysir and continue all the way to locality 3.4 at Grímsnes. The journey takes you across cultivated lowlands, passing by the low mountains *Vörðufell* [64.0686, −20.5428] and *Hestfjall* [64.0127, −20.6686]. Vörðufell is an eroded remnant of the Pliocene–Pleistocene basement rocks projecting through the early Holocene sedimentary cover. Several major strike-slip faults formed in association with earthquake activity in the South Iceland Seismic Zone dissect the mountain and are easily detected from a distance. Hestfjall (1.7 km³) is a volcano of the table-mountain type, formed by an eruption through thin

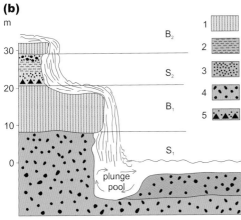

Figure 3.4 (a) Gullfoss waterfall. (b) the stratigraphy at Gullfoss: 1, basalt lava; 2, mudstone; 3, sandstone; 4, conglomerate (orange and green show separate units of conglomerate); 5, tillite; B1 and B2, lower and upper threshold layers; S1 and S2 upper and lower sedimentary units.

89

ice at or very close to a glacier margin during Younger Dryas time (11 700–12 800 years ago) or thereabout. Hestfjall is also in an offrift setting where magma has found a pathway to the surface through faults associated with the South Iceland Seismic Zone (see Fig. 1.2).

Locality 3.4   Grímsnes volcanic field

The *Grímsnes* [64.0420, −20.8866] volcanic field is an inconspicuous group of 12 scoria-cone volcanoes sitting about 15–20 km to the east of the main axis of the West Volcanic Zone and in line with the South Iceland Seismic Zone. The Grímsnes volcanoes were formed in a series of mixed eruptions that took place 8200–9700 years ago and in total produced 0.6 km³ of lava. Each volcano is typically made up of a row of two to four scoria cones delineating short (300–1000 m) volcanic fissures and a small lava flowfield 3.5–23.5 km² in area. The exception is the *Kerhóll* [64.0582, −20.8445] volcano, where a single spatter cone was formed around a circular vent.

The largest edifice of the Grímsnes group is the *Seyðishólar* [64.0658, −20.8426] scoria cone volcano, featuring > 100 m-high cones, whereas the oldest ones are the Selhóll cones. The best-known edifice is undoubtedly the crater *Kerið*, [64.0411, −20.8851] a popular tourist stop that is often mistakenly cited as an example of an explosion crater or a maar volcano because of the small pool retained within it. Kerið is a mixed cone made up of spatter and scoria, which belongs to the Tjarnarhólar cone row. In the eruption some 9000 years ago, Kerið contained a small lava pond that was drained in the final stages of the eruption, leaving behind a deep and steep-sided crater. The crater was deep enough to penetrate the local groundwater table, allowing water to flow freely into the crater to form the small lake.

On the way
In crossing the *River Sog* [64.0049, −20.9730], the road turns along the foothills of *Mt Ingólfsfjall* [63.9764, −21.0252], where the upper half of the mountain (i.e. the top 150 m) is a lava delta formed when the Plio-Pleistocene lava that caps the mountain flowed into the sea. This lava delta rests on a pedestal of a series of significantly older lava flows belonging to the Hreppar Formation. The large boulders and blocks that occur in abundance at the base of the mountain are a stark reminder of the forces that loom beneath your feet, because most of them have fallen from the upper cliffs during major earthquakes in the South Iceland Seismic Zone.

Continuing westwards on Highway 1, the large scarp that appears on approaching the town of Hveragerði is an old seacliff carved out into young volcanic formations by the invading seas at the end of the most recent glaciation. As the winding road climbs up this old seacliff, we re-enter the West Volcanic Zone. On a good day the view from the top of Hellisheiði is a gem. To the east are the Southern Lowlands, with the volcanoes of the East Volcanic Zone, including the Westman Islands, in the background. To the west, Reykjavík appears in the distance, with the Hengill central volcano standing tall in the foreground. Upon descent, the highway crosses a young-looking lava flow, sometimes referred to as Christianity lava (also known as *Svínahraunsbruni* [64.0326, −21.4616]), which is generally believed to be the lava formed by the eruption mentioned when the members of the Alþing were debating the adoption of Christianity in the year 1000 (see p. 80–81).

Hreppar–Þjórsárdalur–Hekla

This excursion is designed such that one can proceed directly from the Gullfoss–Geysir area. From Geysir follow Road 30 to the south towards the town of Flúðir and continue to the intersection with *Road 32* [64.0461, −20.4116] and then turn left (signposted Árnes, Þjórsárdalur and Sprengisandur). This is the starting point of the second excursion.

Here we are practically on the shore of the ancient South Iceland Bay about 100 m above present sea level. Before us are the Southern Lowlands, which were at the bottom of the sea as the Weichselian glacier retreated (see Fig. 3.2). Shortly afterwards, the lowlands emerged from the sea because of the coupled effects of isostatic rise of the land and the construction of vast sandur plains through the accumulation of debris transported by glacial rivers. At *Búðafoss* [64.0328, −20.3293] ('Búði rapids') are the outermost moraines formed during the Younger Dryas stage, where the glacier was calving into the sea some 11 700–12 800 years ago.

On the way. The road from *Árnes* [64.0413, −20.2509] to *Þjórsárdalur* [64.1216, −19.8712] runs parallel to the River *Þjórsá* [64.0603 −20.0755], the longest river in Iceland, stretching some 230 km from its source at Hofsjökull to the southern coast, and which has an average discharge of 400 m³/s. Along this stretch are islets in the middle of river, which are covered by lush vegetation, because their location prevents intervention by man or livestock. South of the river is the Great Þjórsá lava (see p. 77). The mountains and hills north of the road belong to the Hreppar Formation, a

Plio-Pleistocene sequence built mainly of subaerial basaltic lava flows and clastic deposits of fluvial and debris-flow origin. Several distinctive tillite horizons and thick glacial lagoon deposits occur within the upper part of the sequence, indicating the onset of full-scale glaciation about 2.4 million years ago. The Hreppar succession is heavily dissected by faults, and the tectonic fabric is characterized by older north-trending strike-slip faults and younger northeast-trending normal faults.

Locality 3.5
Gaukshöfði–Villingadalsfjall sandur-plain succession

Pull off onto the side road by *Gaukshöfði* [64.0778, −20.0245]. The southern slope of Mt Villingadalsfjall may at first glance look like any other mountain slope in the region, and certainly its geological significance is not revealed in an instant. The lower part of the slope features large grass-covered, terrace-like benches, whereas barren rock is exposed in the steeper upper part. The benches are in fact large blocks that have been displaced downwards by faulting. For example, the rock formations exposed in the cliffs by the main road at Gaukshöfði, which includes an excellent outcrop of pillow lava, are the same formations as occur at the top of Mt Villingadalsfjall, 300 m up.

Walking up the lower half of the slope, these formations are repeated at each crossing of a northeast-trending ravine. The ravines are, in fact, the fault traces. Occasionally a basaltic dyke pokes up in the ravines; the dykes used the fractures created by the faults to rise towards the surface. The benches were formed by a step-wise displacement of the strata along northeast-trending faults and, in all, the cumulative downthrow is nearly 300 m. The faults mark the northern margins of the East Volcanic Zone.

In the upper half of the slope, the original stratigraphy is intact, and it reveals an interesting story. The sequence exhibits distinctive stratification and most of it is composed of fluvial conglomerates and sandstones (Fig. 3.5). Scattered in between these fluvial deposits are lenticular lava bodies with upper surfaces that show clear evidence of having been modified by fluvial erosion. The fluvial deposits were formed by a braided river system emerging from a glacier situated farther inland. In fact it is an old sandur plain, where the debris transported by the ancient rivers and sporadic lava eruption filled in a large palaeo-valley. The western slope of this valley was the east side of Mt Hagafjall, where the móberg ridge that forms the mountain core is blanketed by steeply dipping alluvial-fan deposits.

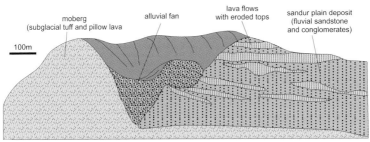

Figure 3.5 Simplified cross section of the geology of Villingadalsfjall.

Locality 3.6 Þjórsárdalur

Þjórsárdalur [64.1216, −19.8712] valley is cut into an extinct central volcano, which bears the same name. This is revealed by an unusual abundance of andesite and rhyolite dyke, lava and tephra formations in the Plio-Pleistocene strata of the surrounding mountains. The valley is, in fact, an old caldera that has been extensively modified by glacial erosion. The volcano centre is at the head of the valley near *Mt Fossalda* [64.20921, −19.74913], where many inclined cone dykes dissect strongly altered and colourful rhyolite and andesite formations.

Sections of the original caldera faults are seen at *Mt Skeljafell* [64.1289, −19.8024] on the east side of the valley and at *Grjótárgljúfur* [64.1884, −19.8838] on the west side, showing that the caldera was at least 8 km wide, rim to rim. At these localities, rhyolite dykes have intruded the caldera faults and connected to small rhyolite lava domes a little higher up in the sequence. At Mt Skeljafell (see Fig. 3.1: site 2) the dyke is exposed in a small ravine north of the main road, where it ascends the western slope of the mountain and can be followed up to the lava dome (the pale rock formation) where it sits on the caldera fault. Among the rocks that form the caldera fill are andesite tuffs and pillow lavas that are closely associated with tillite beds. A good example of this can be seen at *Mt Reykholt* [64.1446, −19.8409] in the centre of the valley, where a thick pillow lava and móberg tuff sequence is sandwiched between two tillite layers. This sequence clearly shows that, about 2 million years ago, while the volcano was still active, a glacier filled in the caldera. Later the volcano was almost completely buried by a series of basaltic lava flows formed by eruptions outside the Þjórsárdalur central volcano, and then its caldera was carved out by erosion during subsequent glaciations, forming the valley we see today.

In the past 7000 years Þjórsárdalur has been repeatedly devastated by tephra fall from the nearby Hekla volcano, but never more so than by the two largest Plinian eruptions (H4 and H3) within the Hekla volcanic system some 4200 and 3100 years ago. The whitish specks on the surrounding mountain slopes are not snow but the remnants of the *pumice-fall* [64.1282, −19.8246] deposit from these Hekla eruptions. At the foothills of Mt Skeljafell this fall deposit is more than 2.5 m thick. When the first settlers arrived in Iceland in the ninth and tenth centuries, the valley had completely recovered from its previous encounters with Hekla, and it soon became quite densely populated, hosting at least 20 farms. This settlement in Þjórsárdalur Valley was decimated by the tephra fall from the 1104 eruption at Hekla. At the time, the valley was completely blanketed by a 10–30 cm-thick pumice-fall deposit, which was enough to make the area uninhabitable. Some of these ancient farm ruins were excavated by the Nordic archaeological expedition in 1939, and one of them is the old Norse homestead at *Stöng* [64.1516, −19.7524] (see Fig. 3.1: site 3).

The valley floor is covered by an ~3000-year-old lava flow – one of the youngest Tungnaár lavas, which originated from volcanic fissures in the Veiðivötn area. It is a branch of a larger lava flowfield, the Búrfell Lava, which covers the high plateau to the east Mt Skeljafell and *Mt Búrfell* [64.0816, −19.8119]. The lava found its way into the Þjórsárdalur Valley through a narrow ravine named *Gjáin* [64.1514, −19.7381], which is about 500 m to the east of the farm ruins of Stöng. Gjáin is a beautiful hollow with two small waterfalls surrounded by grassy slopes, which makes it an ideal place for a picnic on a good day. The geology at Gjáin shows that this small gully has been a feature of the landscape for quite some time, and that the Búrfell lava is not the first one to flow through it (Fig. 3.6).

In the centre of the Þjórsárdalur Valley, the Búrfell lava contains a cluster of small scoria cones, which were formed by rootless eruptions when the lava covered the wetlands that once existed in the valley. At *Hjálparfoss* [64.1144, −19.8535] the River Fossá has eroded a beautiful section through the lava, exposing the roots and conduits of the rootless cones, along with spectacular fanning columns formed by water-enhanced cooling of stagnant lava.

On the way

Travelling from Þjórsárdalur to Hekla the route (Road 32) crosses a pass that the Tungnaár lavas followed down to the lowlands. The major glacial rivers, Þjórsá and Tungnaá, also flow through this pass, and a suite of dams

(a)

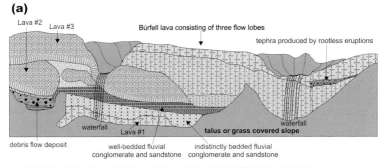

(b)

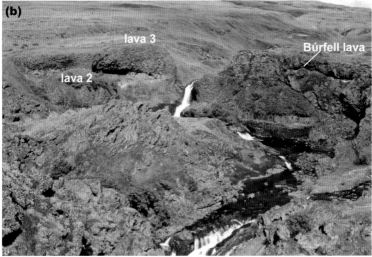

Figure 3.6 (a) Cross section showing the main geological formations at Gjáin, where four separate lava flows (lava 1, 2, 3 and Búrfell lava) fill in pre-existing river channels; **(b)** a photograph of Gjáin that corresponds to the left half of the section.

have been built in this area to serve hydroelectric power plants that produce more than 75 per cent of the country's electrical energy. The route takes you by the *Búrfell* [64.1052, −19.8331] and *Sultartangi* [64.1807, −19.5612] hydroelectric power plants. As you cross the Sultartangi bridge, turn south on to Road 26. Turn east at the intersection with *Landmannaleið* [64.0922, −19.7475] and follow it to the destination at Skjólkvíar. At that point, one should have a good view of the queen, the Hekla volcano.

Locality 3.7 Hekla – the queen of Icelandic volcanoes

Hekla [64.0133, −19.5921] (1491 m) is without doubt Iceland's most famous volcano and is the third most active volcano in the country, behind

Grímsvötn and Katla, with a tally of 18 eruptions in historical times (Fig. 3.7). Hekla is the only one of its kind, because it is the only ridge-shaped stratovolcano known in Iceland where eruptions occur repeatedly on the same fissure (see Figure 3.8). The Hekla edifice is, in the main, constructed by lava and tephra from hybrid andesite to dacite eruption during the last 3000 years or so. In this time it has also featured, albeit rare, rhyolitic explosive eruption and sporadic basalt fissure eruptions have taken place on the associated fissure swarm but outside the central volcano. The nature and location of magma storage beneath Hekla is still poorly constrained, but most likely features a deep-seated (15–20 km depth) and large basaltic reservoir as well as a shallow (3–8 km) and smaller storage zone containing more evolved (dacite to rhyolite) magmas. The andesitic, i.e. the

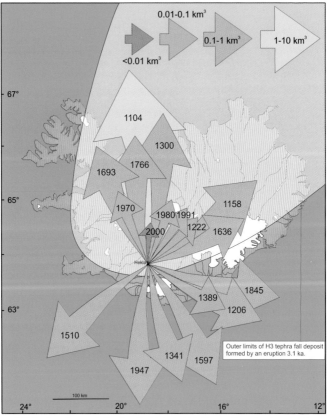

Figure 3.7 Tephra-dispersal direction for historical Hekla eruptions (and years of eruption); the shaded area shows the extent of the H3 tephra layer formed ~3100 years ago by a Plinian eruption at Hekla.

intermediate, magmas, which are so common at Hekla, are most likely produced by mixing of the two above-mentioned end-member magma types.

The Hekla volcanic system has produced a vast number of eruptions in postglacial time, although the three rhyolitic Plinian eruptions H3 (3100 BP), H4 (4200 BP) and H5 (7000 BP) stand out as they blanketed up to two-thirds of the country with pale pumice fall deposit (Fig. 3.7). The first historic eruption from the Hekla volcano was the explosive Plinian eruption of 1104, emitting 2.5 km^3 of rhyolite tephra covering more than half of Iceland with pumice (Table 3.1). More commonly, Hekla events are hybrid eruptions that begin with a shortlived (< 1 hour) explosive phase at the summit crater, producing subplinian to Plinian fall deposit of dacite or andesite composition. Then there is a sharp transition into a fissure eruption, with a curtain of fire extending across the crest of the volcano, producing andesite tephra and lava. This phase typically lasts between several hours and a couple of days, when the activity becomes confined to a short segment of the fissure or a single vent that primarily produces lava.

Hekla erupted four times in the twentieth century – in 1947, 1970, 1980 and 1991 (Fig. 3.8) – and celebrated the turn of the millennium by an eruption on 26 February 2000. Records of other historical Hekla eruptions show that the volcano has erupted at least once every hundred years since 1104, except in the fifteenth century (Fig. 3.7 and Table 3.1). All of these eruptions are mentioned in the historical chronicles, thus facilitating more exact dating of the eruptions than would otherwise be possible. Some eruptions are described in great detail, especially the younger ones (i.e. post-1693), and demonstrate a surprisingly pragmatic attitude towards these fearsome events. This is clearly illustrated by the description of the eruption at Hekla in 1300, written by Einar Hafliðason in the fourteenth century:

> Coming up of fire in Mount Hekla with such violence that the mountain split, so that it will be visible as long as Iceland is inhabited. In that fire, great rocks played like coals and when they hit together there were great crashes that were heard in the north of the country and in many places elsewhere. So much pumice flew onto the homestead at Næfurholt that the roofs of the buildings were burnt off. The wind was from the southeast, and it carried northwards over the country such dense sand between Vatnsskarð and Axarfjarðarheiði, along

Table 3.1 Historical eruptions at Hekla volcano.

Year (AD) of eruption	Tephra (km³)	Tephra (km³) DRE	Lava (km³)	Preceding interval, yrs
2000	0.01	0.004	0.17	9
1991	0.02	0.01	0.15	10
1980–81	0.06	0.026	0.12	10
1970	0.07	0.03	0.2	22
1947–48	0.18	0.08	0.8	101
1845	0.13	0.03	0.63	77
1766–68	0.5	0.20	1.3	73
1693	0.45	0.18	*0.9*	56
1636	0.18	0.08	*0.5*	39
1597	0.29	0.13	*0.9*	86
1510	0.32	0.14	*1.0*	120
1389	0.15	0.07	*0.5*	47
1341	0.18	0.08	*0.5*	40
1300	0.55	0.22	*1.5*	78
1222	0.04	0.02	*0.1*	15
1206	0.4	0.18	*1.2*	46
1158	0.4	0.16	*0.1*	53
1104	1	0.40		>230
Total	**4.93**	**2.04**	**10.57**	

DRE, dense rock equivalent. *Italicised* numbers in column 3 represent anticipated lava flow volumes using the mean ratio of tephra and lava calculated from eruptions where both values are known.

Figure 3.8 Hekla viewed from east-northeast during the 1991 eruption. The dark areas are the lava flows produced during the 1991 event. The white areas are snow-covered slopes of Hekla. The arrow points to the still-fuming vent that remained active for the duration of the eruption.

with such great darkness, that no-one inside or outside could tell whether it was night or day, while it rained the sand down on the earth and so covered all the ground with it. On the following day the sand was so blown about that in some places men could hardly find their way. On these two days people in the north did not dare put to sea on account of the darkness. This happened on 13 July.

The total volume of magma erupted by Hekla in historical times is about 10 km³, which is enough to build a 1 m-wide and 1.6 m-high wall around the 6000 km-long coastline of Iceland (Table 3.1). The tephra fall from Hekla eruptions has repeatedly caused great damage to pasture and cultivated farmlands, with more than 50 farms in the vicinity (< 70 km) of the volcano damaged or destroyed in a single eruption. In more distant regions, the damage felt was different, where much of the livestock was killed by fluorine poisoning through the intake of ash particles while grazing in fields affected by ashfall. The greatest damage was caused by the eruptions of 1300, 1341, 1510, 1693 and 1766.

The composition of the Hekla magma system is roughly a linear function of the repose periods between eruptions. As the repose period increases, the tephra produced by the initial Plinian phase of each eruption becomes more evolved (i.e. contains more silica) and subsequently the explosivity of the initial phase increases. However, the effusive phases that follow always produce lavas of andesite composition (52–58% SiO_2). This compositional pattern in the eruption products indicates chemical stratification of the magma in the shallow storage zone beneath Hekla. Such chemical stratification can be produced by magma mixing, crystal fractionation and assimilation of the surrounding crustal rocks.

The exact nature of this magma storage zone, its location at the west margin of a propagating rift, and its association with a crustal weakness, all contribute to the high eruption frequency of Hekla. In this context it is important to note that eruptions in the immediate vicinity of Hekla, such as the 1554 *Rauðubjallar* [63.9424, −19.7528] (Pálsteinshraun) and 1913 *Lambafit* [64.0739, −19.4042]–*Mundafell* [63.9845, −19.5634] fissure eruptions, produced basaltic-lava flows of magma composition less evolved and distinctly different from that erupted by Hekla summit eruptions. These basalt magmas are most probably derived from a deeper-seated magma storage zone situated below the shallow storage zone of Hekla proper.

On the way
The travel from the *Þjórsá bridge* [63.9297, −20.6649] to *Seljalandsfoss* [63.6162, −19.9939] at Eyjafjallajökull takes you across cultivated sandur

plains constructed by the glacial rivers over the past 11 700 years. The area is, in essence, the marshland of a large delta, which explains why most of the farmhouses stand on hilltops. The low-lying areas used to be covered with peat bogs and fens, but have been drained extensively for cultivation. This region was the stage of Brennu–Njálssaga, one of the best-known of the Icelandic sagas. In the northeast, the ragged points of the Tindfjöll volcano rise to 1460 m, which are all named after trolls and other mythological figures, such as Saxi, Ýmir and Sindri. A large explosive eruption took place in the Torfajökull–Tindfjöll area about 54 000 years ago. This eruption produced widespread tephra fall that is found in the ocean all around Iceland, and also a large ignimbrite sheet that spread southwards over the Þórsmörk area, where it is exposed in accessible outcrops.

Seljalandsfoss at Eyjafjallajökull is an impressive 65 m-high waterfall that cascades off an ancient wavecut cliff and is unique in the sense that it can be viewed from the back as well as the front. An easily accessible footpath passes behind the waterfall, a walk well worth the effort. The grass-covered slopes at Seljalandsfoss are a good vantage point for viewing the vast sandur plain of the Markarfljót River. The farmlands in front of *Fljótshlíð* [63.7348, −20.0214] and Eyjafjallajökull stand on a sandur plain that has been constructed in less than 11 700 years by a 60 km-long braided river system stretching from the Þórsmörk area to the sea. The upper part of the sandur plain fills in a 2–6 km-wide valley, whereas the broader lower section (i.e. the Landeyjar area) is the fanning delta. The main river is *Markarfljót* [63.6175, −20.0156], which originates at *Mt Rauðufossafjöll* [63.995, −19.3761], to the east of Hekla. It flows south in a narrow gorge through the mountainous region between Tindfjöll and Mýrdalsjökull before it enters the sandur plain at Þórsmörk. On its 100 km journey to the coast, the Markarfljót grows steadily as it captures the drainage from its tributaries, which feed off the meltwater from the surrounding glaciers. The total drainage area of the Markarfljót River is 1070 km^2. The discharge of the Markarfljót River is 80 m^3/sec on average, but it fluctuates wildly, commonly with about a tenfold increase in flow during spring thaw or at times of heavy rain. Before the river was confined to its present course by manmade barriers and walls, the frequent fluctuations in the discharge resulted in an unstable course and, with time, the river channels wandered across the plain. In historical times these frequent shifts in the river course have caused severe damage to the farmlands.

Chapter 4

Vestmannaeyjar (the Westman Islands)

General overview

The best way to get to the Westman Island (i.e. Heimaey) is by car/boat or by aeroplane. Travel by car/boat follow Highway #1 east from Reykjavík and turn off to the ferry at Bakkafjara after about 135 km. The turn off is a short distance east of the village Hvolsvöllur. Sailing time to Heimaey is approximately 20 minutes. A flight from the domestic airport at Reykjavík to Heimaey takes about 30 minutes.

In about AD 870 the blood-brothers Ingólfur and Hjörleifur sailed for Iceland with the intention of settling there for good. During the first winter, Ingólfur made camp at Ingólfshöfði; Hjörleifur made his winter camp at Hjörleifshöfði. In the following spring Hjörleifur and his men were killed by his Irish slaves (often referred to by the Vikings as 'Westmen'), who after the event fled to a group of small islands off the south coast. When Ingólfur found out about the fate of his blood-brother, he sailed to the islands and killed all the Westmen. Thenceforth, the archipelago was known as Vestmannaeyjar ('Westman Islands').

The Vestmannaeyjar volcanic system is located on the seaward extension of the East Volcanic Zone and is characterized by alkali magmatism. Approximately 80 volcanic edifices are known within the 900 km^2 covered by the system, of which 18 rise above sea level as skerries and islands (Fig. 4.1). Vestmannaeyjar is a young volcanic system that came to life <100 000 years ago. The islands forming the archipelago are much younger, as they were all formed in the past 10–20 000 years, and most by emergent submarine eruptions similar to the 1963–7 *Surtsey* [63.3015, −20.6038] eruption (see p. 107–108). However, the largest island, *Heimaey* [63.4426,

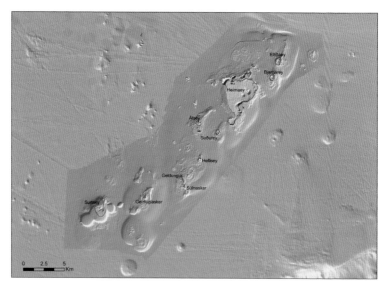

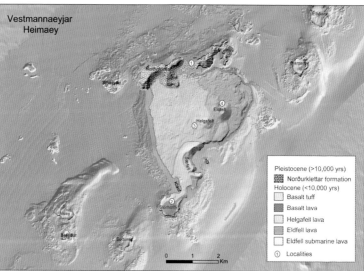

Figure 4.1 (**a**) Overview of the Vestmannaeyjar volcanic system, showing the relative position of Heimaey and Surtsey as well as highlighting the archipelago and the volcanic edifices on the surrounding sea-floor. (**b**) Geological map of Heimaey and the islands in its immediate vicinity.

−20.2750], is a central volcano in the making and is constructed from a tight cluster of basaltic volcanoes that were produced by series of eruptions over the past 15 000 years (Fig. 4.1a). The eruption of more-evolved mugearite to hawaiite magma in the 1973 Eldfell eruptions confirms the status of Heimaey as a juvenile central volcano. Consequently, the geology of Heimaey reveals a more diverse style of volcanism than is found on the other islands, including the products of subaerial basaltic to intermediate effusive as well as explosive magmatic and hydromagmatic eruptions. Studies of the products from the 1963–7 Surtsey and 1973 *Eldfell* [63.4323, −20.2496] eruptions show that the main magma-holding reservoir of the Vestmannaeyjar system is located at 25–30 km depth and that it supplies the system with basaltic magma. The magma erupted in the 1973 Eldfell eruption is derived from the Surtsey magma by fractionation, which occurred in a shallow holding chamber over the ten years that separated the eruptions.

Heimaey

Heimaey, the largest and the only inhabited island of the Vestmannaeyjar archipelago, is 5.5 km long and 4 km wide, with a total area of 13.4 km². It has between four and five thousand inhabitants and is the home of the largest fishing fleet in Iceland. The island is made up of five principal geological formations (Fig. 4.1a), which are described in some detail below.

Locality 4.1   Norðurklettar

Norðurklettar [63.4488, −20.2645] ('northern cliffs') is made up of products from at least seven eruptions, which are in stratigraphic order of Háin, Blátindur, Klif along with Dalfjall, to the west, and Heimaklettur, Miðklettur and Ystiklettur to the east. The Norðurklettar Formation was formed towards the end of the Weichselian glaciation, 13 000–15 000 years ago. At the time the cliffs comprised two separate islands that were moulded into their present shape by wave erosion during the Holocene and formed the isthmus Eiðið. *Háin* [63.4404, −20.2873] offers a spectacular sight into a volcanic vent emerging from the sea, including the outer rims of a tuff cone and an inner scoria cone, representing wet (subaqueous) and dry (subaerial) stages of the eruption.

Locality 4.2   Stórhöfði and Sæfell

Between 7000 and 6500 years ago yet another eruption took place and created the third island, *Stórhöfði*, [63.4007, −20.2868], the southernmost point of Heimaey. Stórhöfði is a typical emerging volcanic edifice, featuring a small lava shield of pāhoehoe capping the tuff cone formed during the emerging phase. The Stórhöfði eruption was similar to that of Surtsey, in that it emerged out of the sea during the stage of explosive activity and then the style of the eruption changed to effusive (lava-producing) when the island had grown large enough to prevent access of the sea to the vents. During the explosive activity in the Stórhöfði event, northerly winds deflected the volcanic plumes towards the south and the subsequent tephra fall built a northward-facing half-cone that forced a northward advance of the lava formed by the effusive activity that followed. The Stórhöfði lava shield is a prime example of a pāhoehoe lava flow field, and numerous lava tubes are visible within the sequence.

About 6200 years ago a submarine explosive eruption began just north of Stórhöfði and formed a 1 km-diameter tuff cone Sæfell, the largest cone structure on Heimaey. The **diatreme** that forms the topmost part of the Sæfell conduit reaches more than 800 m below the present surface, which is remarkable considering its tuff cone origin. *Sæfell* [63.4159, −20.2749] has a volume of 1 km^3 and was thus a sizeable addition to the island of Stórhöfði, but was not large enough to link the three islands into a single landmass. Informative coastal sections into Sæfell can be found at *Klauf* [63.4104, −20.2838] and *Brimurð* [63.4068, −20.2787]. Near Klauf (at [63.4076, −20.2787]) there is a soil horizon situated on the Stórhöfði lava flow but overlain by tuff sequence from Sæfell.

Locality 4.3 Helgafell

Helgafell [63.4292, −20.2601] (rising 227 m above sea level) in the centre of Heimaey is a spatter-scoria cone formed in an eruption about 5900 years ago. It is the first purely magmatic (Strombolian to effusive) eruption within the archipelago that also produced the pāhoehoe lava flow field that now covers the western side of the island. This event marks the birth of the island of Heimaey because the lava it produced connected the Sæfell–Stórhöfði and the Dalfjall–Klifið islands into a single landmass. The lava also closed the seaway south of the Norðurklettar, thus providing the conditions to form the isthmus, Eiðið, which is the natural breakwater between the edifices Klifið and Heimaklettur. Outcrops exposing the contact between Sæfell tuff sequence and the Helgafell lava is present at *Vilhjálmsvík* [63.4112, −20.2864] and *Skarfatangi* [63.4228, −20.2521], although the access to the latter is challenging. The summit of Helgafell provides, on a good day, a beautiful view of the island and the archipelago. In the past, following the invasion of the Algerian pirates in 1627, it was used as a look-out post, and the remains of the shelter used by those on the look-out can still be seen.

Locality 4.4   Eldfell

The most recent addition to Heimaey is the scoria cone Eldfell ('mountain of fire') and its lava flow, both formed in 1973. The onset of the eruption came with little warning. A few mild earthquakes were felt after 10 p.m. on 22 January and the sharpest one occurred in the early morning of the 23rd at 1.40 a.m. At this time only two seismometers were in operation in Iceland, one in Reykjavík and the other in Vík in Mýrdalur, and using these two meters it could not be resolved whether the seismic epicentre was in the Torfajökull volcano or Heimaey. Of these two, Torfajökull was deemed the more likely source, which explains why the eruption appears to have caught everyone off guard. On Monday night, 22 January, people in Heimaey went to bed as usual after a normal workday. The whole fishing fleet was in dock because of a severe storm. Just before 2 a.m. there was a telephone call to the local police station informing the officers on duty that an eruption had started at 200–300 m east of the farm, Kirkjubær (Fig. 4.1a). At first the officer did not take the reporter seriously, but soon it became clear that something unusual was underway, thus police officers drove to the area for further investigation. Upon arrival, they saw an erupting fissure east of Kirkjubær, extending from the harbour mouth in the north and to the sea in Stakkabót, the vent of the Sæfell tuff cone (Fig. 4.1).

The eruption began on Tuesday 23 January at 1.55 a.m. and initially formed a semi-continuous curtain of fires (lava fountains) along the whole length of the fissure. Within three days the activity was reduced to one vent, and it is this vent that built up the cone we now know as Eldfell. Initially, the lava advanced down slope from the fissure, or to the east and the northeast towards the sea. The emergency alarm was sounded in town. The police and the fire brigade drove around town with their sirens on to alert the inhabitants. In about two hours most of the inhabitants were afoot, and as a measure of caution the council decided to evacuate the whole town immediately, apart from a few volunteers who remained to undertake necessary duties. It was quickly realised that there could be imminent danger of the fissure lengthening, either to the north, potentially blocking off the harbour, or to the south, threatening to damage the airport. Furthermore, the mainland civil-defence, domestic airlines and the NATO Defence Force in Keflavík were contacted to provide air transport for hospitalized and elderly people. That night, 300 people, mostly the sick and aged, were airlifted to Reykjavík. The rest of the town's population rushed down to the harbour, having just had time to put on warm clothing and cram a few necessary belongings into a duffle bag. Thanks to the storm from the day before, around 60–70 fishing boats were in dock. They were quickly prepared for departure and the first one left for Þorlákshöfn at 2.30 a.m., followed by a steady stream of fishing vessels packed with people. The whole operation went remarkably smoothly and without mishaps, thanks to the decisive response from local authorities, favourable weather conditions and a little luck. In the first 12 hours the eruption spewed forth more than 30 million tonnes of tephra and lava, which were initially dispersed to the north and east.

The eruption lasted just over five months until 3 July 1973. It covered most of the town with several metres of tephra, and about a quarter of the town was buried by the lava. An extensive network of pipes and pumps was installed to spray the lava with seawater. In total 6.2 million tons of seawater were sprayed on the lava, and this action was successful in slowing down the advance and redirecting the flow. However, despite ample efforts, parts of the town could not be saved and 417 houses got buried under lava and volcanic ash. Remains of some houses can still be seen in the lava field. The remains of the old swimming pool are visible at [63.4424, −20.2598], a house is protruding from the lava at [[63.4410, −20.2646], and the old electric cable poles are still standing at [63.4433, −20.2607]. A project nicknamed 'Pompei of the North' began excavating some of the houses buried in the tephra.

The summit of the Eldfell cone provides a good overview of the lava field. From there one can easily hike to the so-called *Páskahraun* branch of the lava [63.4300, −20.2401], which advanced into the sea during Easter 1973, as well as Vagabönd, which is a scoria mound representing a section of the Eldfell cone that broke off on 19 February and was then carried by the lava flow towards the sea. *Kirkjubæjarhraun*[63.4393, −20.2430], the bulldozed sector of the lava flow field, is the site where it is thickest, or 70–120 m. In this area the lava was harnessed for its heat. This was achieved by spraying the lava with seawater and collecting the steam generated for space-heating in the town over a period of 25 years or until 1998. A rather peculiar lighthouse can be found at [63.4365, −20.2278] as it is mounted on wheels, so it can be relocated due to the rapid erosion of the coastline.

The size of Heimaey increased by 2.2 km² (20%) in the 1973 eruption and the Eldfell eruption expelled about 0.25 km³ of magma onto the surface, compared to the 1.2 km³ erupted by the 1963–7 Surtsey eruption, and produced a lava flow field covering 3.3 km². The eruption and its impact on the island is nicely illustrated at the Eldheimar museum (http://eldheimar.is/en/).

The 1963–7 Surtsey eruption

An island emerging from the sea

Surtsey [63.3015, −20.6038] is a small volcanic island situated about 33 km off the central-south coast of Iceland and is the westernmost island of the Vestmannaeyjar archipelago. The island is the subaerial part of the larger Surtsey volcano, a 6 km-long east-northeast-trending submarine ridge that rises from a depth of 125 m and covers about 14 km² (Fig. 4.2). The prominent features on Surtsey are two abutting 140 m-high tuff cones and a small pāhoehoe lava flowfield that caps the southern half of the island.

Surtsey is also the youngest of the Vestmannaeyjar, formed by a prolonged eruption that was first noticed on 14 November 1963 by fishermen attending their nets about 20 km southwest of Heimaey. The eruption intensified and gradually built an island that in its prime rose to 170 m

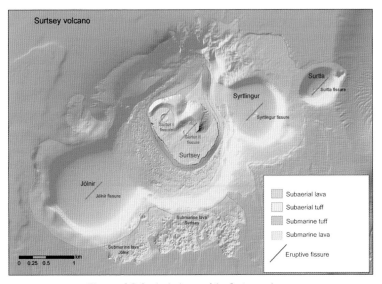

Figure 4.2 Geological map of the Surtsey volcano.

above sea level. It was named Surtsey after Surtur, the fire-raising giant of Norse mythology who was to set fire to the Earth on Judgement Day. Surtsey rumbled and lava flowed for 3½ years until June 1967, when Surtur called it quits – the longest eruption in Iceland since settlement was over and, for the first time, Icelanders had witnessed the formation of a submarine table mountain.

The tuff cones, Surtur I and II (Fig. 4.2), were built by Surtseyan explosive eruptions over a period of five months between November 1963 and April 1964. During this stage the eruption columns rose as high as 13 km. When the island had grown large enough to seal the volcanic conduit from the invading seawater, the activity changed from being wholly explosive to purely effusive. The eruption featured two prolonged subaerial effusive eruption phases, the first erupting through the Surtur II crater and lasting for 13½ months (April 1964 to May 1965). It produced a 100 m-high lava shield that covered 1.53 km^2 above sea level. In the second phase the Surtur I crater came to life again to produce a 70 m-high and 1 km-wide lava shield between August 1966 and June 1967. The total volume of tephra and lava produced by the Surtsey eruption amounts to about 1.2 km^3, of which about 0.4 km^3 is lava. Today, erosion is the main force at work on Surtsey and, since the end of the eruption in 1967, wave erosion has reduced the surface area of the island by 1 km^2.

Long before the eruption stopped, the island was proclaimed a nature reserve and all visits to it were restricted, so biologists could make the most of this unique opportunity to study how life would migrate to and develop on this new and desolate land. In 1964 the first living organism, found in the ash deposits closest to shore, was a tomato plant, and in May that same year a single fly was found on the island. Seagulls hung out in the tidewaters farthest from the crater and the first vascular plant to flower on the island was the sea rocket, found along the shore in 1965. Today five species of birds nest on Surtsey: herring gull, black-backed gull, black guillemot, kittiwake and fulmar. The last species was the first one to nest and hatch its young successfully on Surtsey in 1970.

Chapter 5

The central south

General overview

Central-South Iceland extends from the Markarfljót River in the west to Jökulsá at Breiðamerkursandur in the east (Fig. 5.1). In the west it stretches across the East Volcanic Zone along the foothills of the glacier-capped *Eyjafjallajökull* [63.6296, −19.6368] and *Mýrdalsjökull* [63.6328, −19.0545] volcanoes, and then follows the margins of the zone to the northeast before turning east towards the *Öræfajökull* [63.9886, −16.6465] volcano across the vast and desolate plains of *Skeiðarársandur* [63.9485, −17.4128]. In the north the largest glacier in Europe, *Vatnajökull* [64.3943, −16.9779], towers over the region. Currently, this is the most volcanically active area in Iceland, as can be deduced from district names such as *Eldsveitirnar* [63.6845, −18.2833] ('Fire Districts') and *Öræfi* [63.9611, −16.8673] ('wasteland'). Over the past millennium it has experienced not only more but also larger eruptions than any other part of the country, as we have been reminded, not once but three times in the first 11 years of the twenty-first century.

However, although destruction and hardship are brought on by the volcanic eruptions, the fertility of the soil is greatly enhanced by the tephra falls they produce, and the towering volcanic mountains generate clouds and precipitation. Consequently, the area enjoys the highest rainfall in Iceland, and in between the rugged mountains are valleys with fertile pastures. This is the only region in Iceland where the land is painted green with vegetation from the sea to the highest mountaintops. Even the talus slopes are grass covered. As the soil cover creeps down the steep talus slopes, it often forms majestic stairways that in places appear to stretch up into the sky above.

We present two excursions across Central-South Iceland. The first covers the coastal regions from the Markarfljót River to Mýrdalssandur. The second is centred on the Fire Districts, the region between the large

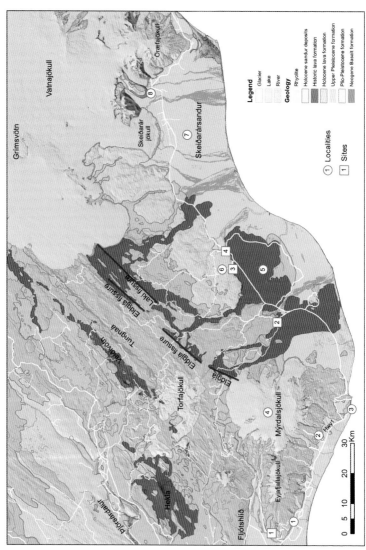

Figure 5.1 The main geological features of Central-South Iceland.

sandur plains of *Mýrdalssandur* [63.4795, −18.6827] and *Skeiðarársandur* [63.9485,−17.4128], but also includes the regions around the national park at *Skaftafell* [64.0270, −16.9949].

Eyjafjallajökull and Mýrdalur
The icecaps and the secluded volcanoes

This excursion starts at the *Markarfljót bridge* [63.6163, −20.0151] (Highway 1) and begins by examining several aspects of the Eyjafjallajökull volcano. From there it tracks the southern flanks of the Mýrdalsjökull volcano as we cross the plains of *Skógasandur* [63.5039, −19.4831] and *Sólheimasandur* [63.4735, −19.3761] into the valley *Mýrdalur* [63.4471, −19.1120] and on to the western part of *Mýrdalssandur* [63.4795, −18.6827].

Eyjafjallajökull volcano

Eyjafjallajökull [63.6296, −19.6368] is a 1666 m-high shield volcano, as is suggested by its gently sloping east–west profile (Table 1.4), and is situated near the tip of the propagating Eastern Volcanic Zone. It is crowned by a small caldera and an icecap that reaches down to 1000 m above sea level. The south slope of the volcano terminates in old wave-cut sea cliffs that rise to 500 m above present sea level, whereas the north side has been strongly modified by glacial erosion (Fig. 5.2). The Eyjafjallajökull volcano has been active for at least 800 000 years and sits on more than 25 km-thick crust. It consists of at least six glacial and six interglacial volcanic sequences. The former sequence comprises subglacial volcanic formations, such as pillow lavas, kubbaberg breccia, hydroclastic tuffs and intercalated tillites. The latter sequence is typically composed of a stack of subaerial lava flows indicating emplacement during ice-free periods. Good examples of these sequences are exposed in the sea-cliffs on the south side.

Compared to its neighbour volcano, *Katla* [63.6328, −19.0545], Eyjafjallajökull volcano has been relatively quiet in postglacial times. While Katla has produced >400 eruptions over the last 11 000 years, Eyjafjallajökull has recorded between 14 and 16 eruptions in the same period, including the infamous 2010 event. Three events are known on the western flanks of the volcano; the Hamragarðar and Kambagil basalt lava flows were formed by two early Holocene fissure eruptions, while the third is represented by a substantial cone row a little further up on the western slope (Fig. 5.3a). Six to eight small lava-producing early Holocene events are known at the

Figure 5.2 Aerial view of Eyjafjallajökull and Mýrdalsjökull volcanoes showing the difference in their profile and the dimensions of their summit calderas. View is to the east. South and north margins of Eyjafjallajökull caldera are indicated by single-headed arrow, while that of Mýrdalsjökull is indicated by double-headed arrow. Also indicated is the approximate position of the Katla fissure.

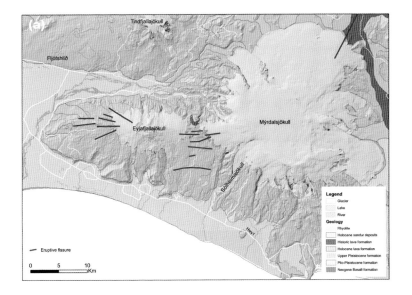

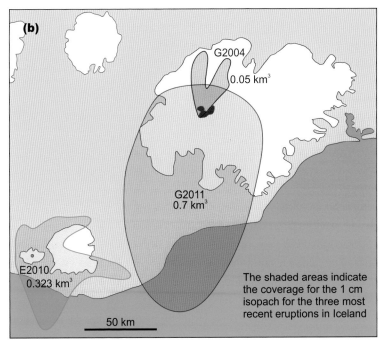

Figure 5.3 (a) Map of Eyjafjallajökull, showing Holocene fissures and volcanic formations. (b) Outline of the 1 cm isopach for the tephra fall deposits of Grímsvötn 2004, Eyjafjallajökull 2010 and Grímsvötn 2011 eruptions to illustrate the difference in magnitude and intensity of these events.

Fimmvörðuháls [63.6260, −19.4443] pass on the east side of the volcano. A total of five events are known from the upper reaches of the volcano. Four of these, AD 500, AD 1613, AD 1821–23 and AD 2010 took place through the summit crater, while one was produced by a radial fissure on the volcano's northwest flank (Fig. 5.3).

The 2010 Eyjafjallajökull eruptions

In May 1994 a seismic swarm took place beneath the volcano, lasting almost for one month. This was the first seismic swarm at Eyjafjallajökull detected by the Icelandic SIL seismic monitoring network, installed in 1988, and related to magma injection into the crust beneath Eyjafjallajökull. An event was not unexpected because of the known activity in historical time. Five years later, in July 1999, another seismic swarm was again detected beneath Eyjafjallajökull. Similarly to the previous one, it lasted for about a month. These events resulted from intrusion of sills emplaced at the contact between the volcano's base and the crust it rests on. The centre of surface deformation produced by these intrusions, as evaluated via satellite-based interferometry, was just southeast of the summit crater, and it was evident that Eyjafjallajökull was getting ready for eruption. Nonetheless, almost ten years passed before the volcano released the next episode of unrest. In May 2009 a swarm of earthquakes was accompanied by a significant inflation of the outer flanks of the volcano. Yet again, unrest stopped before an eruption. However, this time around, only six months passed before the volcano showed its restlessness again. In late December seismicity as well as deformation (= inflation) resumed, increasing steadily in magnitude and intensity over the following three months. On 4 March the rate of inflation sped up and the earthquake centres began to migrate towards the surface. On 18 March it was evident that Eyjafjallajökull was heading towards an eruption and the Icelandic Civil Defence raised the alert. At 11:30 p.m. on 20 March, a 400 m-long fissure opened up on the northeast flanks of Eyjafjallajökull, just above the mountain pass Fimmvörðuháls (Fig. 5.3a). The pass is the saddle between the volcanoes of Eyjafjallajökull and Mýrdalsjökull. A curtain of fire with up to 150 m-high lava fountains emerged from the fissure, feeding an 'ā'a lava flow of basaltic composition, which advanced over snow without melting it before cascading into two of the gullies that dissect the pedestal of Eyjafjallajökull above Þórsmörk. The flank eruption above Fimmvörðuháls lasted until 12 April and produced 0.026 km³ lava and tephra.

The pause did not last, because in the early morning of 14 April elevated seismicity and tremor showed that an eruption was imminent. This time the magma of intermediate composition (i.e. trachyandesite) broke out from vents at the summit of Eyjafjallajökull and through 250 m of ice. Consequently, debris-laden meltwater cascaded down the outlet glacier *Gígjökull* [63.6627, −19.6211] and destroyed the glacier lagoon at its front (Fig. 5.4). From noon and throughout the afternoon the eruption discharged floodwater down Gígjökull to such a degree that the sandur plain of the Markarfljót River was completely flooded, including the segments of Highway 1 on either side of the Markarfljót bridge. All through the day white, steam-rich plume was lofted into the atmosphere from the erupting vents at the summit (Fig. 5.5a). In the early hours of the evening or around 7 p.m., the appearance of the eruption plume changed in an instant. Its colour became distinctly dark, almost black, because now the vents were vigorously spewing tephra (ash, lapilli and bombs) into the atmosphere, reaching altitudes up to 8 km (Fig. 5.5b). From this time onwards until 18 April, the eruption featured distinct explosive periods, where each period

Figure 5.4 Gígjökull lagoon before and after the 2010 eruption. Before the eruption the glacier-carved bowl was filled with pristine glacial waters, but as shown by the lower panel, now it is filled to the brim with volcanic debris. Inset shows the lagoon on 15 April, the second day of the eruption

Figure 5.5 Snapshots of the activity during the 2010 summit eruption: **(a)** Steam plume rising from the summit region early on 14 April. (b) Tephra-laden plumes produced by explosive activity around 7 pm on 14 April (first phase). (c) Weak explosive activity accompanied by lava emission on 22 April (second phase). (d) Eruption plume from explosive activity of renewed intensity on 5 May.

lasted for two to three hours and was followed by half hour to hourly breaks. Individual explosive periods were typified by pulsating activity with explosions repeated on the timescale of seconds. This first phase of the eruption peaked in the evening of 14 April with southeast dispersion of the eruption plume and then again on 17 April, when the plume was directed southwards by strong wind (Fig 5.3b). In both instances the ash-rich plumes reached the British Isles and the continent, forcing a widespread closure of the airspace over Europe, resulting in major disruption to air travel and transport for almost 10 days (15 to 24 April). The intensity of the eruption dropped significantly on 18 April (Fig. 5.5c), and a few days later it cleared the path for re-opening of the European airspace. The explosive activity became sporadic and produced weak plumes rising only 1–4 km into the atmosphere. At the same time lava began to pour out from the summit vents and advanced down the path of the outlet glacier Gígjökull. This second phase of the activity lasted until 4 May. On that day, an intense earthquake swarm was detected deep under the volcano or at depths of 10–13 km. This seismic activity was accompanied by inflation of the volcano, indicating that a new, fresh magma batch was migrating up through the volcano plumbing system. On the following day more intense explosive activity resumed, producing tephra-laden eruption plumes reaching maximum heights of 6 km (Fig. 5.5d). Once again north-westerly winds carried the ash-rich plumes in over Europe, forcing partial closure of the airspace on 4–8 May and again on 16–17 May. This third eruption phase lasted until 17 May, and after that activity dwindled steadily until stopping on 22 May. Minor phreatic (steam-driven) explosions took place in early June. The summit eruption extruded about 300 million m^3 of tephra, of which about 50% was very fine ash (i.e. grains < 63 micrometres in diameter) and 23 million m^3 of lava. Thus, the total volume of magma produced by the summit event was 323 million m^3, which is enough material to blanket Greater London with 20 cm of ash.

Locality 5.1 Hvammsmúli

At the base of *Hvammsmúli* [63.5758, −19.8715], a small promontory south from Eyjafjallajökull, outcrops a 45 m-thick massive pale-grey body of highly olivine-pyroxene phyric ultrabasic rock known as ankaramite, which represents the most primitive rock type of the transitional alkali series. The 0.6 million-year-old Hvammsmúli ankaramite is best exposed at *Dysjarhóll* [63.5744, −19.8754], Pöst and in a nearby quarry, and is directly overlain by a 2–4 m-thick unit of bedded tephra and thin lava of the same composition (Fig. 5.6). Originally, the Hvammsmúli ankaramite was

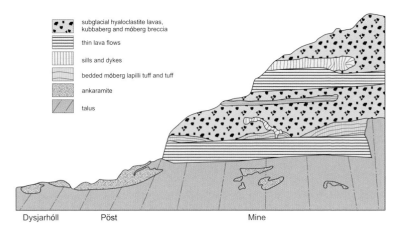

subglacial hyaloclastite lavas, kubbaberg and móberg breccia

thin lava flows

sills and dykes

bedded móberg lapilli tuff and tuff

ankaramite

talus

Dysjarhóll   Pöst   Mine

Figure 5.6 A cross section showing the lithostratigraphy at Hvammsmúli.

interpreted as a sill, but new investigations suggest that it may represent a lava lake formed in association with the overlying ankaramite tephra and lava. The upper part of Hvammsmúli consists of subaerial olivine-phyric basalt-lava flow sequences alternating with thicker units of subglacial hyaloclastite breccias and tuffs.

On the way

The *Skógafoss* [63.5312, −19.5124] waterfall by the old settlement farmstead Skógar is perhaps the most perfectly shaped waterfall in Iceland, fallingover a 60 m-high cliff tucked in between bright green grass-covered slopes.

Locality 5.2   Hofsá River – myth and the truth about the formation of Skógasandur and Sólheimasandur

The Book of Settlement tells the story of a dispute between Þrasi at *Skógar* [63.5279, −19.4995] and Loðmundur at *Sólheimar* [63.4946, −19.3281], two farms on either side of Jökulsá at *Sólheimasandur* [63.4735, −19.3761]. According to the folktale, one morning Þrasi saw a great flood approaching from the mountains and used sorcery to force it into a more easterly path towards Sólheimar. Loðmundur, who was blind, was told by one of his slaves that an ocean of water was cascading down the mountain slopes and heading their way. Loðmundur knew Þrasi's game, so he used his own powers to redirect the floodwaters westwards in the direction of Skógar. Þrasi responded and Loðmundur repeated his act. This apparently went on for a while until they agreed on a floodpath midway between the farms, where the *River Jökulsá on Sólheimasandur* [63.4984, −19.3989] now runs directly to the sea. Also, according to one version of the tale, these floods formed the sandur plain of Sólheimasandur.

For a long time it was thought that the floods mentioned in the tale of Þrasi and Loðmundur had formed the whole outwash plain of *Skógasandur* [63.5039, −19.4831] and Sólheimasandur, because geological evidence appeared to suggest an age within historical times for the main tephra-laden flood deposits on both sandur plains. However, recent findings show that the main flood deposits at Skógasandur are prehistoric (Fig. 5.7) and were formed by a jökulhlaup that accompanied a seventh-century Katla eruption. In this eruption the vents were in the western part of the caldera and within the drainage area of Sólheimajökull outlet glacier. However, younger flood deposits of lesser volume that formed during Settlement age cover part of Sólheimasandur. We also know from tephrochronological studies that the early settlers soon became acutely aware of the awesome nature of Katla, because within 70 years of the arrival of the first settlers it had erupted three times. After being exposed to two relatively small Katla eruptions in the late ninth and early tenth centuries, the newcomers witnessed the largest flood-lava eruption on Earth in the past 2000 years, the

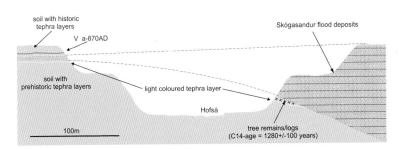

Figure 5.7 Schematic cross section of the Hofsá River [63.5237, −19.4552] channel where it dissects the main Skógasandur flood deposits. On the east bank of the river the tephra-rich flood deposits are several metres thick and overlay a soil containing many prehistoric tephra layers and a distinct horizon with tree-remains near the top. The ^{14}C age obtained from these tree remains is 1280±100 years (i.e. about AD 600). In the west bank the flood deposits are much thinner (0.5 m) and overlain by soil containing historical tephra layers including the Settlement layer formed around 870. Therefore, the Skógasandur floods occurred well before the first settlers arrived in Iceland.

AD 934–40 Eldgjá–Katla Fires. One of these eruptions is likely to have been responsible for the historical jökulhlaups onto Sólheimasandur, which supposedly kindled the dispute between Þrasi and Loðmundur.

Locality 5.3 Dyrhólaey and the islands in the sand: volcanoes telling the tale of a different past

Some of the unusual features of the region around Mýrdalur are the mountains *Pétursey* [63.4673, −19.2711], *Dyrhólaey* [63.4075, −19.1137] and *Hjörleifshöfði* [63.4166, −18.7551], which stand as isolated crags in a sea of sand and generally a good distance inland from the current coastline. Now on dry land, these structures are remnants of emergent submarine volcanoes formed when these low-lying coastal plains were fully submerged. Since then, the rivers have transported enough debris to raise the area out of the sea and link the former islands with the main landmass of Iceland.

Dyrhólaey (Portland) is the second southernmost point of Iceland (63°23′N), after Kötlutangi, and is a heavily eroded submarine volcano of the Surtseyan type (Fig. 5.8). The main part of Dyrhólaey is built of well-bedded tuff formed by hydromagmatic explosions, and it represents the remains of a larger tuff cone. On the eastern flank of Dyrhólaey, the tuff sequence is capped by compound pāhoehoe lava, which in places exhibits cube jointing, indicating water-enhanced cooling of the lava. The lava flows represent the subaerial phase of the eruption when the tuff cone had grown large enough to prevent seawater from accessing the vent. Flowing away from the vent(s), the lava re-entered the sea and cooled rapidly to form the cube-jointed lobes. The Dyrhólaey sequence is typical for Surtseyan eruptions and, if visibility allows, the type-volcano Surtsey can be seen from this location as the southernmost island of the Westman Islands.

Other, perhaps a little more obscure, evidence of submarine volcanic activity is found above Mýrdalur in the móberg mountain above the *Skammidalur* [63.4512, −19.1005] farm. The mountain is a remnant of a submarine tuff cone made up of bedded lapilli tuff containing isolated blocks of fossiliferous marine sandstones and conglomerates. The fossil-bearing blocks, up to 1 m in diameter, were plucked from the underlying substrate by the Surtseyan explosions in the volcano conduit and incorporated as accidental clasts (i.e. xenoliths) in the tuffs.

Most of the fossils are molluscs, with 15 species of bivalves (lamellbranchia) and 9 species of snails (gastropoda), but they also include species

(a)

(b)

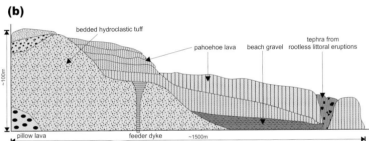

Figure 5.8 (a) Aerial view of Dyrhólaey; (b) Simplified cross section showing the structure of the Dyrhólaey tuff cone (~100 m deep, ~1500 m wide).

of brachiopods, echinoderms (sea urchin), crabs (crustacea) and worms. Although the Skammidalur fossil assemblage bears a strong resemblance to modern marine fauna around Iceland, almost half are warm-ocean species that now reside farther south in the Atlantic or are extinct. Thus, when these fossil animals were living off the coast of Iceland, the ocean was considerably warmer than it is today. The age of the Skammidalur shell-bearing blocks is 2–3 million years (early Quaternary), as suggested by the similarity of the fauna to that found in the *Serripes* layers at Tjörnes in North Iceland (see p. 175).

Similar shell-bearing blocks are found sporadically in the móberg tuffs extending from Pétursey, which is a 113 ka old volcano, in the west to

Höfðabrekka [63.4267, –18.8982] in the east, indicating that the volcanic formations of Mýrdalur rest on an extensive sequence of marine sediments that also covers the submerged bedrock shelf to the south, well beyond the Westman Islands. A progressive southward younging of the sediments is suggested by the fossil-bearing sandstone xenoliths ejected by the explosive eruptions at Surtsey in 1963–4, which contain fossils of only modern day species. Thus, some 2–3 million years ago the sea inundated the Mýrdalur area, and it is likely that the now-majestic volcanoes of Mýrdalsjökull and Eyjafjallajökull began their life some 600 000–700 000 years ago in a similar fashion to the Westman Islands, as a cluster of small volcanic islands in the middle of the sea.

Locality 5.4
Mýrdalsjökull volcano, the Katla fissure and Mýrdalssandur
The Mýrdalsjökull central volcano (1450 m) is not only the second largest volcano in Iceland but it has also produced some of the most cataclysmic eruptions in the country's history (Fig. 5.2). Second to the Hofsjökull volcano, with a total volume of 380 km³, it is capped by Iceland's fourth largest glacier (580 km²). Mýrdalsjökull volcano has been the locus of activity on the 100 km-long Katla volcanic system over the past 200,000 years and is characterized by basalt magmatism (see Table 1.4). However, it has produced more evolved magmas in significant quantities, as is evident from scattered occurrences of rhyolite lavas and ignimbrites on the volcano, and dacite tephra layers in the soils of the surrounding lowlands.

The volcano is crowned by a large (~110 km²) ice-filled summit caldera that hosts the legendary *Kötlugjá* [63.6076, – 19.0220] ('Katla fissure'). According to an old folktale, the name 'Kötlugjá' is derived from a cook at the monastery at Þykkvabæjarklaustur, named Katla – a woman true to ancient traditions and the proud owner of magic trousers. The trousers allowed anyone who put them on to run as far and as long as they desired without getting tired. One autumn the shepherd Barði wore the priceless trousers, without permission, when gathering his sheep. When Katla discovered this, she secretly pulled Barði aside and drowned him by the main door in a large barrel of whey. Over the winter, the whey was gradually used up by the residents of the monastery, and Katla realized that eventually her crime would be exposed. As Barði's body was about to emerge, Katla put on her magic trousers, ran onto the Mýrdalsjökull and flung herself into a chasm in the glacier. Shortly thereafter, the chasm erupted, sending a colossal flood (jökulhlaup) in the direction of the monastery. Accordingly, it was believed that Katla had used her witchcraft in an attempt to avenge her fate and, ever since, the chasm has been known as Kötlugjá. Consequently, it has become common practice to refer to the caldera at Mýrdalsjökull central volcano as Katla, and the outbursts that occur within its boundaries as Katla eruption

It is not surprising that the activity at Mýrdalsjökull has been linked to the wicked trickery of the supernatural, considering the frequency of Katla eruptions and the hardship and devastation they caused in the surrounding farming communities. With more than 20 eruptions in historical times, Katla is the second most active and the most productive volcano in Iceland (Fig. 5.9; Table 5.1). It is notorious for its violent eruptions, caused by explosive interaction between the ice and the fire, as the magma tunnels its way through the 400 m-thick glacier. Hence, Katla eruptions are typically hydromagmatic and are accompanied by widespread tephra fall and colossal jökulhlaups consisting of meltwater, ice and volcanic debris. The most recent eruption of Katla was in 1918, which lasted for 24 days or from 12 October to 4 November (Table 5.1).

At about 1 p.m. on 12 October a sharp earthquake was felt in Mýrdalur, and two hours later an eruption column up to 16 km high was seen rising from the Katla fissure, dispersing tephra towards the north-northeast in such quantities that it laid to ruin the farm (*Búlandssel* [63.7832, −18.5833]) in the Skaftártunga district. At about the same time the first jökulhlaup was seen to burst forwards in two places from the outlet glacier Kötlujökull, to the west of *Hafursey* [63.5189, −18.7738] and at *Kriki* [63.6238, −18.8682]. This flood covered much of the sandur plain. The jökulhlaup had two main flood paths. At the Hafursey outlet the flood was directed towards the south, following the Múlakvísl and Sandá river beds before cascading onto the sea on either side of Hjörleifshöfði. The Kriki outlet directed the floodwaters eastwards across the plain into the rivers Hólmsá, Skálm and Landbrotsá and then along the course of the Kúðafljót River to the sea (Fig. 5.10).

A second flood pulse emerged from the Hafursey outlet at 5 p.m., transporting so much ice that it 'looked as if snow-covered hills were rushing forwards'. The floodwaters advanced at velocities of 15–20 km/hour, with a peak discharge of around 200 000 m^3/s. The total volume of the 1918 jökulhlaup is estimated at 3–5 km^3 and it flooded more than 50% of the Mýrdalssandur, covering 400 km^2. In the process it dumped 0.5–1 km^3 of

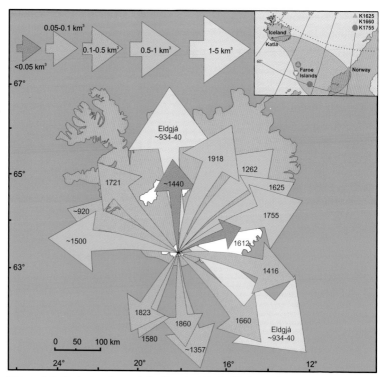

Figure 5.9 The tephra dispersal directions for historical Katla eruptions. The arrows indicate main dispersal direction for each eruption, arrow size is proportional to size of eruption (as indicated in key) and the numbers indicate year of eruption. Inset shows localities where tephra fall from historic Katla eruptions was observed outside of Iceland.

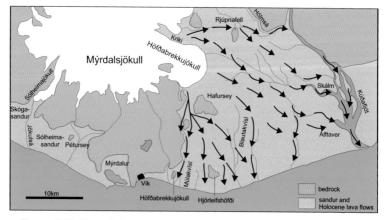

Figure 5.10 The main flow paths of the jökulhlaup formed by the 1918 Katla eruption.

Table 5.1 Historical eruptions of the Katla system

Eruption site	Eruption year/ century	Month	Date	Length (days)	Tephra volume (km3)	Repose period
Katla	1955	June	25			37
Katla	1918	October	12	24	1.00	58
Katla	1860	May	8	20		37
Katla	1823	June	26	28		68
Katla	1755	October	17	120	1.35	34
Katla	1721	May	11	100	0.33	61
Katla	1660	November	3	60	0.26	35
Katla	1625	September	2	13	1.25	13
Katla	1612	October	12		0.04	32
Katla	1580	August	11			80
Katla	1500				0.50	
Katla	15th century					
Katla	1440					24
Katla	1416				0.3	59
Katla	1357				0.20	95
Katla	1262				0.48	17
Katla	1245					66
Katla	1179					
Katla	12th century					
Eldgjá	934–940				4.50	14
Katla	920				0.27	
Katla	9th century					

tephra and other debris onto the sandur plain, elevating its surface by about 1 m. The 1918 jökulhlaup also added 14 km² of new land to the coastline in front of Hjörleifshöfði. This new land was called *Kötlutangi* [63.3985, −18.7456] and became the southernmost point of Iceland. Kötlutangi remains its honorary status despite extensive erosion of the unconsolidated flood deposits by the prevailing coastal currents.

Katla continued to erupt explosively for the next 23 days, although at decreasing intensity, dispersing tephra to the northwest and southeast of the volcano. The total volume of tephra from the 1918 eruption is estimated at about 1 km³, roughly equal to 0.4 km³ of magma.

Since the twelfth century the eruptions at Mýrdalsjökull have been centred on the Katla fissure (Table 5.1) and have progressed in much the same manner as the 1918 eruption described above. These events were all explosive subglacial eruptions producing tephra falls and were accompanied by jökulhlaups onto Mýrdalssandur. However, the magnitude of these eruptions has varied greatly. The smallest, such as those of 1823 and 1860, produced less than 0.02 km³ of tephra and were accompanied by jökulhlaups of small to moderate magnitude. However, the largest, namely the 1262, 1625, 1721, 1755 and 1918 eruptions, emitted large quantities of tephra (0.5–1.5 km³), with tephra fall extending to more than 200 km from source and large jökulhlaups of magnitude similar to or larger than the one in 1918. These explosive basaltic eruptions included events of Plinian intensities, with eruption columns rising to altitudes of 23–27 km and were powerful enough to cause tephra fall in Scandinavia and Western Europe (Fig. 5.9, insert).

By the same token, the magnitude of the damage inflicted by these eruptions on the land and nearby farming communities was significantly greater. For example, over 50 farms in the Fire Districts were abandoned for months and even years in the wake of the tephra fall from the 1755 Katla eruption.

Katla jökulhlaups have caused not only hardship, by destroying farms, cultivated fields and fine pastures in and around Mýrdalssandur, but also significant changes in the environment. In settlement times a coastal lagoon named Kerlingarfjörður, large enough for Viking ships to dock, existed in front of Hjörleifshöfði (Fig. 5.11). The lagoon was destroyed by the first post-settlement jökulhlaup onto Mýrdalssandur, 'Höfðárhlaup' in 1179. The lagoon Kúðafjörður in *Álftaver* [63.5216, −18.3782], another useful harbour, met a similar fate in the so-called 'Sturluhlaup', which occurred as a result of the 1262 Katla eruption. In 2011 there was a small jökulhlaup from Katla, which destroyed the Highway 1 bridge across the river

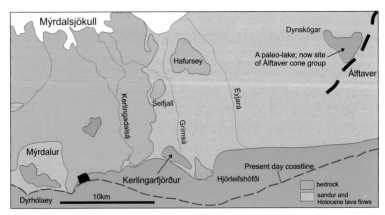

Figure 5.11 Reconstruction of Mýrdalssandur as it may have looked in the 10th Century showing the position of Kerlingarfjörður, which most likely was a coastal lagoon rather than a fjord, as implied by the name (fjörður = fjord).

Múlakvísl. In short, Mýrdalssandur and the neighbouring areas have changed more in the past 1100 years than, for example, the landscape of Western Australia has in 40 million years.

A simple count of Katla tephra layers in prehistoric soils around the volcano reveals that over 400 basaltic and at least 12 felsic explosive subglacial eruptions have occurred at the Mýrdalsjökull volcano in the past 11 000 years. In addition, 5–10 small effusive eruptions have occurred along the northern periphery of Mýrdalsjökull, and the fissure system extending to the northwest has produced two flood-lava eruptions: the 6600 BC Hólmsá Fires and the AD 934–40 Eldgjá Fires. The latter is a giant among historical lava-producing eruptions on Earth. In total the volcano has produced 30–35 km^3 of tephra and close to 25 km^3 of lava in the past 7000 years, making it the most productive volcano in Iceland.

At least two large silicic Plinian, eruptions at Katla took place in Allerød time around 12 000 years BP. In these events more than 10 km^3 of rhyolite rushed out of the volcano, sending pyroclastic flows cascading down the southern flanks and dispersing airborne tephra well beyond the shores of Iceland. The pyroclastic flows are known as the Sólheimar ignimbrite and they outcrop at several localities in Mýrdalur. The tephra fall deposit from these events forms an important time marker in marine sediments in the North Atlantic, where it is known as Ash Zone 1. On land in Iceland this tephra fall is known as the Skógar tephra (see p. 198), whereas in other Nordic countries it is referred to as the Vedde tephra, after the lake deposit in which it was first found in Norway. It is probable that the Katla caldera was formed or substantially enlarged by this eruption. The eruption plumes produced by this eruption dispersed ash over much of the North Atlantic and parts of Europe and may have reached as far as 3500 km from its source.

The current repose period in activity of Mýrdalsjökull volcano (i.e. since 1918) is among the longest known in historical times (Table 5.1). Seismic unrest does occur from time to time and, as a precautionary measure, the road traffic across the sandur plain is then halted on both sides of the plain. Increased activity within the caldera of Katla volcano began in 1999.

The Fire Districts
Skaftártunga to Öræfi
This excursion more or less follows Highway 1 through the farming districts of *Skaftártunga* [63.6939, −18.4989], *Síða* [63.8395, −17.9661] and

Fljótshverfi [63.9142, −17.7169], then across the Skeiðarársandur outwash plain to Öræfajökull.

Nowhere in Iceland are the awesome qualities of volcanic eruptions more evident than in the Fire Districts. The area is guarded by the Katla volcano in the west, and hidden under ice in the northwest is *Grímsvötn* [64.4035, −17.3416], the most active volcano on Iceland. When these volcanoes erupt, they not only spew ash everywhere, but they are also accompanied by violent jökulhlaups that temporarily put hundreds of square kilometres of land under water and destroy everything in their path. The region was also the centre stage for the three largest eruptions since the settlement of Iceland: the basaltic fissure eruptions of *Eldgjá* [63.9628, −18.6157] (AD 934–40) and *Laki* [64.0700, −18.2387] (AD 1783–4), and the Plinian eruption of the Öræfajökull volcano in AD 1362. All of these eruptions altered the landscape significantly and had severe consequences for the inhabitants at the time.

Entering the Fire Districts from the west takes you across the desert-like plain of Mýrdalssandur and then through a cluster of low mounds, which are the Álftaver cone group, formed by the rootless eruptions when the western branch of the Eldgjá lava flows covered the wetlands around the old *Álftaver* [63.5216, −18.3782] settlement in the fourth decade of the tenth century. At this point, a distinct scarp will appear on the horizon to the northeast, marking a line of an old sea cliff dating back to late glacial times. At the eastern end, the cliffs are crowned by *Mt Lómagnúpur* [63.9639, −17.5036], the highest freestanding cliff face in Iceland, rising to 767 m above sea level (see also Fig. 5.16).

The flat-lying area in front of the scarp is an old sea floor that, over the past 15 000 years, has gradually emerged from the sea through the combined effort of glacial rivers and basaltic volcanism. Debris dumped by glacial rivers and their jökulhlaups, along with the periodic advance of large basaltic lava flows, have built up the sandur plains. In the process they have added a 110 km-long and 20 km-wide strip to Iceland, almost in a geological instant.

Farthest east is the glacier-covered Öræfajökull volcano, rising to 2110 m at Hvannadalshnjúkur, Iceland's highest peak. Between Öræfajökull and Lómagnúpur, three outlet glaciers extend from the largest icecap in Europe, Vatnajökull, onto the sandur plain and supply melt-water to the many rivers of Skeiðarársandur. The largest outlet glacier, Skeiðarárjökull, is also the main route for jökulhlaups from the subglacial caldera lake of the Grímsvötn volcano. As a consequence of the eruption at Gjálp, the last eruption-related jokulhlaup was in 1996, when a subglacial fissure erupted through the ice just to the north of Grímsvötn (see p. 140). Since then, small (<0.6 km³) jokulhlaup have occurred every 3–5 years, the last ones in 2015 and 2018.

The age of the basement rocks is 0.1–4 million years and they belong to the Upper Pleistocene and the Plio-Pleistocene formations. The oldest rocks exposed at the foothill of Öræfajökull in the east and west belong to the basement, which is Neogene to early Pleistocene in age (1.7 to 3.2 Ma). The Öræfajökull volcano rests unconformably on this basement and is a significantly younger structure (<0.7 ka). The basement rocks are predominantly volcanic in origin, but also feature some well exposed glacial and lacustrine sediments. The volcanic rocks principally consist of alternating layers of basaltic lava (grey) and móberg (brown) that are often intercalated with tillites and other types of diamictites. The abundance of móberg tuffs and breccias is highest in the west within the younger sequences, and gradually decreases eastwards. This occurrence simply reflects the increase in the frequency of eruptions interacting with ice and glacier meltwater as the influence of the Ice Age glaciers became more dominant. The landscape changes accordingly, from relatively smooth rolling topography with irregular stratification, to steep-sided and regularly stratified mountains separated by deeply incised glacial valleys.

The basement rocks are partly covered by basaltic lava flows produced by Holocene fissure volcanoes and by the extensive sandur plains that are currently undergoing rapid change due to changing climate: lagoons are forming and the braided nature of the rivers is disappearing. In other areas the basement rocks are covered by wind-lain soil or loess, which is extremely fertile. Nowhere in Iceland is the soil thicker than here, up to 12 m in the Skaftártunga district. The soils contain hundreds of tephra layers and thus provide an excellent record of the volcanic history in the region over the past 8500 years. This is well illustrated by the soil profile at *Hólmsárbrú* [63.6354, −18.5199] (see Fig. 5.1: site 2), a stop well worthwhile for those interested in tephrochronology.

Locality 5.5   Historical flood lava eruptions

The Fire Districts were the scene of the two largest flood-lava eruptions on Earth in the past millennium. These eruptions produced the Eldgjá and

Laki lava flows, which together cover 1100 km², an area approaching the size of Greater London (Fig. 5.12). The fact that these two eruptions produced more than half of the magma volume erupted in Iceland over the past 1200 years is a good indicator of their enormous size. These eruptions were truly extraordinary and of catastrophic proportions, as is keenly hinted at by Charles Lyell in his monumental work, *Principles of Geology.*

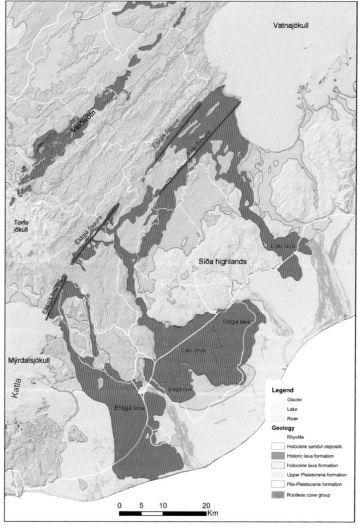

Figure 5.12 The volcanic fissures and lava flow fields of the AD 934–40 Eldgjá and 1783–84 Laki eruptions. Locality 5.5 refers to the general area covered by the Eldgjá and Laki lavas.

Therefore, it is worth considering these eruptions in some detail. Site 3 (*Fjaðrárgljúfur*) [63.7730, −18.1744]) on Figure 5.1 provides a good vantage point for viewing the vast lava fields of these two eruptions.

The 934–40 Eldgjá eruption

The Eldgjá vents form a discontinuous 75 km-long volcanic fissure extending from the Katla volcano in the west, almost to the tip of Vatnajökull in the east (Fig. 5.12). The eruption takes its name from a spectacular 150 m-deep and 8 km-long chasm called Eldgjá ('fire fissure') that occupies the central part of the vent system. However, Eldgjá is an old tectonic graben that was reactivated in the 934–40 eruption. The thick móberg sequence that makes up the walls of Eldgjá is capped by red and black scoria and spatter deposits containing discontinuous lava-like units formed as fountain-fed flows. The scoria and spatter cones at the bottom of Eldgjá are also considered to be primary volcanic features.

The early settlers must have been amazed by the incessant effusion of lava from the Eldgjá fissures, which spread out in two branches that in the end covered about 800 km² (Table 5.2). The eruption also featured more than 16 explosive subplinian to Plinian phases and may have lasted for six years. The explosive activity generated eruption plumes in the range of 11–17 km in height and completely blanketed southern Iceland with tephra within the 0.5 cm isopach, which extends up to 70 km from the vents, and in the area of Álftaversafréttur ('Álftaver highland pastures') it is commonly in the range of 0.5–4 m thick in measured soil profiles (Fig. 5.13). The Eldgjá tephra was dispersed beyond the shores of Iceland and has been detected in the Greenland ice cores among other sites. The magnitude of the eruption is best understood by spreading the Eldgjá lavas along a 35 m-wide and 540 km-long strip, equal to the distance from Paris to Berlin; the lava would form a 1000 m-high wall.

The effects of the eruption on the early settlement in Iceland must have been devastating. Scarce records show that the Eldgjá lava flow advanced

Table 5.2 Some basic facts on the 934-40 Eldgjá and 1783-84 Laki eruptions: mt = million tonnes.

Eruption	Duration	Eruption episodes	Column height	Volume (tephra)	Volume (lava)	Area (lava)	length (lava)	SO2 output
Eldgjá	3-6 years?	15	11–17 km	1.4 km³	18.1 km³	781 km²	70 km	219 mt
Laki	8 months	10	7–13 km	0.4 km³	14.7 km³	599 km²	65 km	122 mt

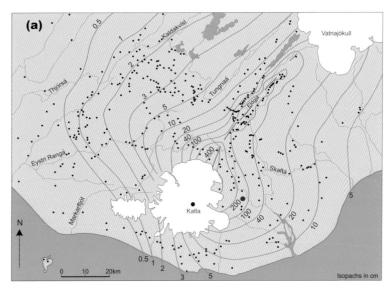

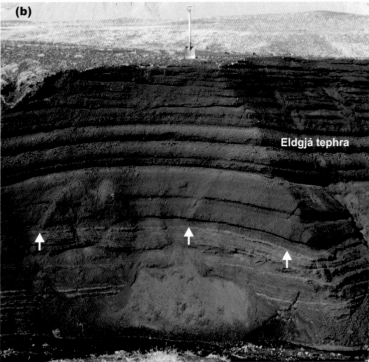

Figure 5.13 (a) Eldgjá eruption tephra dispersal map. Black dots indicate sites of thickness measurements. (b) Eldgjá tephra in a section at Atlaey (red dot on a). White arrows point to base of the Eldgjá tephra and shovel is about 1 m tall.

over vast areas of productive farmland and forced many settlers from their lands, especially within the districts of Álftaver and Síða. Some families were forced to relocate because of thick tephra coverage; others took desperate measures to rake the tephra off their fields. We also know that the lava forced the Skaftá River into a new course closer to the foot of the Síða scarp, just south of its present channel, which follows the margins of the younger Laki lava flow (see Fig. 5.12). Thus, the local environmental effects of the eruption were enormous. The many rootless eruptions, which formed the cone groups of Álftaver and Landbrot more than 40 km away from the actual vents, must have amazed the onlookers, as the lava converted marshy areas and shallow coastal inlets into dry land.

The Eldgjá eruption was the greatest source of volcanic pollution in recent history, exceeding the other two 'biggies', the 1783 Laki and 1815 Tambora eruptions, by a factor of two. It pumped 219 million tonnes of SO_2 into the atmosphere and this gas reacted with atmospheric vapour to produce about 450 million tonnes of H_2SO_4 aerosols and part of it spread to remote regions to cover much of the Northern Hemisphere. In comparison, the most climatically significant eruption of the twentieth century, the 1991 eruption at Mt Pinatubo in the Philippines, injected about 10 million tonnes of SO_2 into the atmosphere. Scanty information in historical accounts, underpinned by climate modelling studies on the potential effects of Eldgjá, indicates that the eruption had a significant effect on the atmosphere and weather patterns in Europe and the Middle East for several years. However, the intensity of the climatic impact of Eldgjá is not thought to have surpassed that of Laki or Tambora, because the eruption was prolonged, and subsequently the sulphur emissions were drawn out over several years.

Good outcrops through the Eldgjá lava are scarce, but are found along the Hólmsá River gorge, at Alviðruhamrar, and in the soil section at Hólmsárbrú, where the lava rests directly on its tephra layer (see Fig. 5.1: site 2). In the thick soil bank along the Skaftá River (see Fig. 5.1: site 4) between the *Ytri-Dalbær* [63.7682, −18.1214] and *Kirkjubæjarklaustur* [63.7895, −18.0528] farms, all tephra layers older than Eldgjá are disturbed, having been slanted and even rolled up into folds. However, the younger historical tephra layers in the top of the soil are undisturbed and form sub-horizontal continuous layers. Here the great Eldgjá lava forced its way under the soil cover and, through advance and lava inflation, it disrupted the soil. The historical tephra layers were formed after this event and are therefore undisturbed.

The *Landbrotshólar* [63.7597, −17.9708] cone group was formed by rootless eruptions when the Eldgjá lava advanced onto the marshy coastal plains in front of the Síða scarp. It is by far the largest cone group in Iceland and at present it covers an area of 60 km^2. Part of the group is now buried beneath the 1783 Laki lava flow and it may have originally covered 150 km^2. In 1793 Sveinn Pálsson examined the Landbrotshólar cone group and described it as pyroclastic cones with a lava cap (i.e. welded spatter). He was the first to suggest that it was formed by secondary eruptions.

The 1783–84 Laki eruption

We know more about the Laki flood lava eruption than any other of its kind, because it is described in many contemporary accounts, including several chronicles that contain enough details about the eruption to be regarded as Eldrit ('chronicle of an earth fire'). However, none are as spectacular as the Eldrit, 'A complete description of the Síða Fires', written in 1788 by the Reverend Jón Steingrímsson at Prestbakki in the Síða district. This chronicle not only contains vivid and accurate descriptions of the events of the eruption on a day-to-day basis and its consequences for the local community, but also demonstrates how intrepid we can be when faced with the extremes of Mother Nature.

In 1783 the people of South Iceland had enjoyed a favourable spring and were looking forward to summer. However, their destiny was about to change. Weak earthquakes in the Skaftártunga district in mid-May were the first sign of what was to come. The intensity of these earthquakes increased steadily and on 1 June they were strong enough to be felt across the region from Mýrdalur and Öræfi. The earthquake activity escalated until 8 June, when a dark volcanic cloud spread over the district, blanketing the ground with ash (Fig. 5.14a). The Great Laki eruption had begun.

Later that day, lava columns more than 1000 m high rose from a new volcanic fissure up in the highlands to the north. The volcanic fumes filled up the atmosphere such that the Sun appeared red as blood and was deprived of its natural brightness. The accompanying rainfall contained salty and sulphurous water, causing smarting in the eyes and on the skin.

On 11 June the channel of the mighty Skaftá River dried up and on the following day a large lava stream surged out from the Skaftá River gorge with an

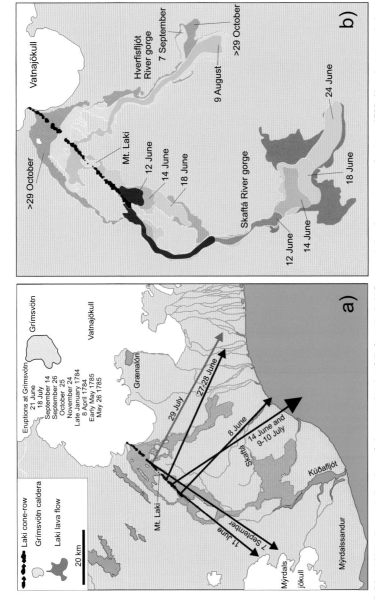

Figure 5.14 (a) Main dispersal directions of plumes and tephra falls from the Laki fissures during the summer and autumn 1783. Also shown are timing of eruptions at the Grimsvötn volcano; (b) growth of the Laki lava flow field reconstructed using information from the contemporary accounts on the position of the lava flow front at the given dates.

incredibly loud cracking noise and rumbles. The volume of this lava was similar to that of the river before it dried up. Recurring tephra falls occurred through June and July, and the lava continued to surge out of the gorge for the next 45 days (Fig. 5.14b). By the time it stopped, the lava had filled up the gorge, which had been up to 100 m deep before the eruption, and covered 350 km^2 of the land, including the cultivated areas in front of the gorge and 17 farms.

However, the Laki fissures were not done. On 29 July the heavy tephra fall from the fissures covered the eastern part of the Fire Districts, forcing many farmers off their land (Fig. 5.14a). On 3 August the channel of the Hverfisfljót River dried up and shortly thereafter fresh lava surged out from its gorge (Fig. 5.14b). One can only imagine what went through people's minds when they saw this new flow advancing and threatening to engulf the Síða district completely in lava.

In the autumn the Fire Districts were showered by intermittent tephra falls and the lava continued to flow from the Hverfisfljót River gorge until the end of October, destroying an additional four farms and adding 250 km^2 to the Laki lava flow field. The Laki fissures continued to produce lava until 7 February, when the fires at the fissures were finally extinguished.

As if one erupting volcano was not enough, in mid-July 1783 'a sandy and muddy ash' fell in the Fire Districts 'from another eldgjá'. This was the first of many concurrent subglacial eruptions at the Grímsvötn volcano, which occurred intermittently until May 1785 (Fig. 5.14a).

The Laki eruption (8 June 1783 to 7 February 1784) nearly exceeded its neighbour Eldgjá, forming the second-largest basaltic lava flow in historical times (Table 5.1). The eruption occurred on a 27 km-long fissure delineated by more than 140 vents and craters (see also Fig. 1.13c). Ten eruption episodes occurred during the first five months of activity at Laki, each featuring short-lived (2–4 days) explosive eruptions followed by a longer-lasting phase of lava emissions. The explosive activity from the eruption produced a tephra layer that covered more than 8000 km^2, the volume of which exceeds that produced by all four Hekla eruptions in the twentieth century. The noxious fumes emitted by the eruption stunted the growth of grass and killed more than half of the livestock through fluorine poisoning. These consequences ultimately resulted in the disastrous 'Haze Famine' that killed 20 per cent of the population.

The effects of the Laki eruption extended well beyond Iceland. Its columns pumped 100 million tonnes of SO_2 into the westerly jet stream, which governs the atmospheric circulation above Iceland, producing sulphur-rich plumes that were dispersed eastwards over the Eurasian continent and north into the Arctic. The sulphur dioxide reacted with atmospheric vapour to produce 150–200 million tonnes of sulphuric acid (H_2SO_4) aerosols, of which about 85 per cent were removed in the summer and autumn of 1783 by subsiding air masses within high-pressure systems (Fig. 5.15). This removal formed the infamous dry fog that hung over the Northern Hemisphere for more than five months. The acidity of the dry fog was such that it caused considerable damage to vegetation and crops all over Europe and stunted tree-growth in Scandinavia and Alaska. Furthermore, the eruption had significant impact on the atmosphere. It produced a temporary change in movement of atmospheric currents and shifted the monsoons, with severe consequences for North Africa, India and Japan. The following winter was one of the most severe on record in both Europe and North America. The eighteenth-century temperature records from these regions suggest that the annual cooling that followed the Laki eruption was about −1.3 °C and lasted for two to three years.

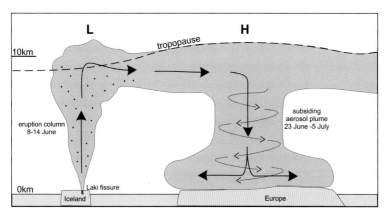

Figure 5.15 Simplified vertical section from Iceland and across mainland Europe showing the eastward transport of the sulphuric aerosol plumes from Laki.

Locality 5.6   The Quaternary Síða Group

The Plio-Pleistocene strata of the Síða and Fljótshverfi districts that form the old sea cliffs above Kirkjubæjarklaustur consist of a 700 m-thick volcanic succession called the Síða Group. It comprises at least 14 stratigraphical packages referred to as Standard Depositional Unit (SDU; described below). The Síða Group unconformably overlies an eroded sequence of flat-lying lavas and sedimentary rocks (the Fljótshverfi Group), formed before the Ice Age set in. The distribution of individual flows appears to have been mainly controlled by the existing topography, as is illustrated by the units that make up *Mt Keldunúpur* [63.82535, −17.98222] and Mt Lómagnúpur in Fljótshverfi (Fig. 5.16).

The lower part of each SDU consists of columnar jointed lava (5–15 m thick) that sometimes is underlain by pillow lava. Placed on top of the columnar jointed lava is 5–40 m thick cube-jointed lava, which in turn is overlain by 10–120 m thick hyaloclastite (kubbaberg breccia and massive lapilli tuff; Fig. 5.16). The SDU is typically capped by a 15 m thick unit of well-bedded lapilli tuff. The lava at the base of the SDU often exhibits spectacular columnar joints, such as those forming the pavement-like Kirkjugólf ('church floor') at *Kirkjubæjarklaustur* [63.7895, −18.0528] and the readily accessible cliffs of *Dverghamrar* [63.8506, −17.8602] below the Foss farm. The cube-jointed lava typically features poorly developed columns with irregular orientation, although locally these fan out to form regular rosette-jointed bodies. The SDUs are often separated by 5–50 m-thick sedimentary units consisting of diamictites, sandstones and mudstones.

In the past 10 000 years the rivers have cut many gorges and gullies into the Síða Highlands, but none as spectacular as the Fjaðrárgljúfur gorge. Here the rather innocent-looking Fjaðrá River has carved out a 100 m-deep ravine into the hyaloclastite flows and other formations of the Síða. In early postglacial times a sizeable lake occupied the valley above Fjaðrárgljúfur (the flat-top benches along the valley side are the **strandlines** formed by that lake) and was kept at bay by an erosion-resistant threshold. At the time, the gorge did not exist and a much larger river flowed down the gently sloping hills before cascading off the cliffs of the Síða scarp. By shear power of flow the river rapidly carved out the gorge back into the mountains. Eventually it breached the threshold at the mouth of the valley, and in so doing drained the lake, changing the conditions to those of today.

Mt Lómagnúpur (767 m) features the highest cliff face in Iceland, with a vertical rise of more than 600 m (Fig. 5.16). It forms the southern end of the Björninn mountain complex, which is built of Pleistocene móberg and lavas. The cliff itself was carved out by glaciers and later modified by coastal erosion. At the base of Lómagnúpur in the west are deposits from two rock

Figure 5.16 Southern face of Mt Lómagnúpur. The upper part of the cliff, above dotted line, provides excellent exposures into the Síða Formation, which here is comprised of several Standard Depositional Units (SDUs), displaying the characteristic internal stratigraphy of the SDU.

avalanches, one of unknown prehistoric age and another that was formed by an avalanche in July 1789; the latter occurs as a hummocky terrain just inside the southern tip of the mountain. Highway 1 traverses the outermost lobes. Ebenezer Henderson examined this rock avalanche in travels around Iceland in 1814–15 and gave the following description:

> Turning round the extremity of the mountain [Mt Lómagnú-pur], which hangs almost directly overhead, at the angle of two ranges of ancient buildings we fell in with many heaps of stones and immensely huge masses of tuffa [móberg], which have been severed from the mountain, and hurled down into the plain, during the rockings occasioned by an earthquake in 1789.

It is known from other sources that this rock avalanche occurred in the early hours of the day, sometime in July 1789, and the earthquake mentioned by Henderson was probably induced by the avalanche rather than being its cause. The scar left behind by the avalanche is clearly visible near the cliff top, but on the ground the avalanche deposit forms a 650 m-wide fan. Approximately 2.5 million m³ (or about 6 million tonnes) of material were displaced in the event.

Locality 5.7 Skeiðarársandur – volcanism in Vatnajökull and the glacial outwash plains

Skeiðarársandur (1300 km²) is the largest sandur plain in Iceland. It is the alluvial fan of Skeiðarárjökull, the largest outlet glacier on the south side of Vatnajökull icecap, and has an ablation area of 1200 km². *Skeiðarársandur* [63.9485, −17.4128] features a complex braided river system that can broadly be divided into two main systems: the *Núpsvötn–Súla* [63.9558, −17.4671] system and the *Gýgjukvísl–Skeiðará* [63.9394, −17.3655] system. The former emerges from the western side of Skeiðarárjökull, and the latter from the eastern side. The rivers Gýgjukvísl and Skeiðará used to flow across the sandur plain in separate river channels. In July 2009 the Skeiðará river channel dried up, as the river had merged with Gýgjukvísl due to the retreat of *Skeiðarárjökull* [64.0182, −17.2439] outlet glacier. The effect of this change is best visualized where Highway 1 crosses the current river channel of *Skeiðará* [63.97579, −17.00029]. The large bridge that is now closed was built in 1974 to provide a crossing over a then much larger

Skeiðará river. This bridge was replaced by the current one in 2017. The size difference is a clear indicator of the magnitude of the change in the 43 years that passed between the construction of the two bridges. Although these river systems relentlessly carry large quantities of debris onto the sandur plain, their contribution to its construction is subsidiary when compared to the transport of material by the catastrophic jökulhlaups that rush the sandur plain at regular intervals. The jökulhlaups at Skeiðarársandur originate in two ways: by sudden emptying of the ice-dammed marginal Lake Grænalón and by draining of the caldera lake at Grímsvötn. The contribution of both of these agents has been reduced significantly due to warming of the climate.

Grænalón [64.1687, −17.3605] is an ice-dammed marginal lake that occupies an ice-free tributary valley on the west side of Skeiðarárjökull (see Fig. 5.12a). The lake is fed by surface runoff from the surrounding mountains and a small outlet glacier originating near *Þórðarhyrna* [64.2684, −17.5457]. When the water in Grænalón reached a critical level (i.e. 90% of the glacial thickness) it had enough momentum to overcome the ice barrier. The water penetrated under the ice, and subglacial tunnels were carved out by the frictional heat generated by the flowing water. This resulted in abrupt draining of the lake and sudden floods in the rivers Súla and Núpsvötn. The ice-dammed lake of Grænalón used to drain at fairly regular intervals or every one to two years. It no longer does so, because decreasing size of outlet glaciers and reduction in glacial runoff due to warming climate has reduced Grænalón to a fraction (<1%) of its former size in and around 1970.

In chronicles from the sixteenth and seventeenth centuries, Grímsvötn ('lakes of Grímnir') are said to be lakes located within the glacier Grímsvatnajökull and possess the peculiar nature of expelling 'Earth fires' (i.e. volcanic eruptions) that bring about large jökulhlaups onto Skeiðarársandur. Grímnir (i.e. one with a mask) is the name that Óðinn (Woden), the chief heathen god in Norse mythology, used when he mingled with humans. Hence, it seems logical that Grímsvötn take their name after masked Óðinn, because after all, the lakes are masked by ice. Grímsvatnajökull is believed to be the original name of the largest glacier in Europe, which we now know as Vatnajökull. In the eighteenth and nineteenth centuries the knowledge of the position of these erupting lakes was uncertain. It was not until they were rediscovered in the early twentieth century that their exact location was firmly established.

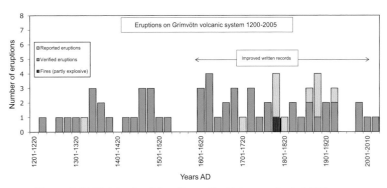

Figure 5.17 Plot showing Grímsvötn eruption history over the last 1200 years.

Grímsvötn is a subglacial lake that occupies the summit caldera of the most active central volcano in Iceland, which bears the same name. It is located at the head of the outlet glacier Skeiðarárjökull (see Fig. 5.14a). It is the most active volcano in Iceland in terms of eruption frequency, with more than 70 basaltic explosive eruptions in the last 1000 years and 65 confirmed events since AD 1200 (Fig. 5.17). Most of these eruptions are thought to have taken place within the summit caldera of the Grímsvötn volcano, although as many as 10 to 11 eruptions may have taken place below thicker ice cover outside the caldera, as was the case with the 1996 Gjálp eruption. Tephra volumes are difficult to verify due to large fall area within the icecap, but an estimated range is 0.01 to 1 km³. Depending on factors such as the ice thickness over the eruption site, only a fraction of the erupted material is dispersed away from the vents through the atmosphere in many eruptions. The best estimates for the volume of erupted magma in individual events have been obtained by using the mass of ice melted by the magma erupted in the twentieth-century events. These calculations indicating average volumes of ~0.04 km³ within the Grímsvötn caldera and ~0.3 km³ for eruptions outside of the caldera. Jökulhlaup from the subglacial lake in the Grímsvötn caldera can be triggered by ice melting associated with eruptions on the Grímsvötn system, as occurred in 1996 when Gjálp erupted. Sometimes the reverse is true; jökul-hlaup trigger eruptions within the Grímsvötn caldera, as occurred in 2004.

The 2004 eruption

This explosive hydromagmatic basalt eruption began in the late evening on 1 November 2004 and lasted for five days. It erupted in sequence from two separate vents located within an ice cauldron in the southwest corner of the

Figure 5.18 Snapshot of the 2004 Grímsvötn eruption: Source vent viewed from the south–southwest at 16:11 on 2 November. At this time vent 1 (west) is erupting tephra while vent 2 (east) is mostly emitting steam.

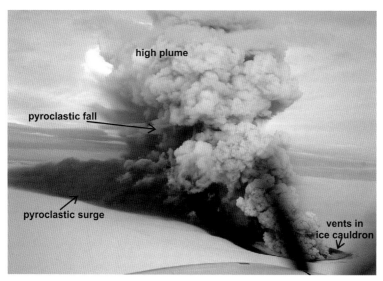

Figure 5.19 The 2004 eruption site at 15:34 on 2 November, showing simultaneous deposition via tephra fall from the high plume and via flow from a pyroclastic surge. View is to the northeast.

Grímsvötn caldera (Fig. 5.18). Explosive activity dispersing tephra north-wards and beyond the confines of the ice cauldron was limited to the first 45 hours of the eruption (Fig 5.3b). An interesting aspect of this eruption

Figure 5.20 Snapshot of the 2004 Grímsvötn eruption: Grímsvötn 2004 tephra deposit (3 m thick) in section at site about 1 km from the source vents. Note the diffuse appearance of the bedding.

was near-continuous fallout from a strongly bent-over, 6–10 km high plume coupled with deposition from recurring pyroclastic surges (Fig. 5.19). In 45 hours this activity had produced a ~11 m thick accumulation of tephra onto the ice surface within 3 km of the source vents. This tephra sequence is typified by diffuse bedding on the decimetre scale, where indistinctly planar to cross-bedded tuff lenses alternate with basalt pumice bearing lapilli tuff or lapilli lenses (Fig. 5.20). The total volume of the tephra produced by the 2004 event is about ~0.05 km³, which is, as indicated above, almost identical to the average volume for an eruption within the Grímsvötn caldera.

The 2011 eruption

After 5.5 years of recharging, as indicated by steady inflation of the volcano since December 2004, an abrupt and intense seismic swarm began underneath Grímsvötn volcano at 5:50 p.m. on 21 May 2011. Magma was moving rapidly towards the surface and by 7:00 p.m. an eruption had broken out at the 2004 vent site (Fig. 5.21a). The sky was clear at the time and soon it was evident that this event was of different magnitude from its immediate predecessors (Fig. 5.21b). The eruption column rose out of the vents at high speeds and reached heights of about 20 km in less than 30 minutes to form a 50–60 km-wide umbrella cloud within one hour of the eruption onset. At much lower altitudes, or at heights of 4–10 km (Fig 5.21b), a dark

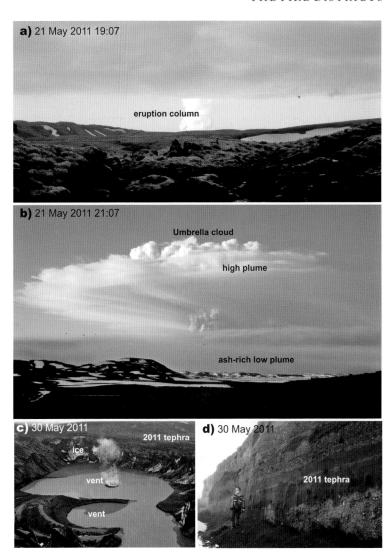

Figure 5.21 Snapshots of the 2011 Grímsvötn eruption: (a) Steam-rich plume rising from source vent on 21 May at 19:07 or seven minutes after the start of the eruption. View is to east. (b) Towering eruption column and associated umbrella just over two hours after the start of the eruption. The height of the plume is about 20 km. View is to the east. (c) The 2011 source vent two days after the eruption ended. (d) Grímsvötn 2011 tephra deposit (4–5 m thick) in section at site less than 1 km from the source vents. Note the thick and very coarse-grained basalt pumice beds.

tephra-laden plume extended southwards from the main eruption column, dispersing tephra over the eastern sector of the Fire Districts. Over the next five hours the area between Skaftártunga in the west to Öræfi in the east was covered by this lower tephra-laden plume such that visibility was down to 1–3 m. Beneath it the land surface was blanketed by tephra more than 1 cm thick (Fig. 5.3b). By six in the morning of 22 May the altitude of the high plume had dropped to 15 km and by noon its height was down to 10 km. Nonetheless, the high plume continued to support the southward-moving, ash-laden lower plume and accompanying tephra fall. On the following day the high plume hovered at 5–9 km altitudes and effectively had merged with the lower plume. In following days the activity dwindled steadily and by 28 May the eruption was over. This was the largest eruption at the Grímsvötn volcano since 1873 (Fig. 5.21c and d). In one week the eruption produced about 0.7 km^3 of tephra and most of it in the first three days. This is more than twice the magma volume produced by 63 days of activity at Eyjafjallajökull in 2010 and more than 10 times the volume produced by the five-day-long Grímsvötn eruption in 2004. Hence, the 2011 event was greater in both magnitude and intensity than the 2004 Grímsvötn and 2010 Eyjafjallajökull events. This is well illustrated by the thickness of the tephra layers at equivalent distance from the source vents. At 7 km distance, the tephra from the 2004 Grímsvötn event is 7 cm thick, 2010 Eyjafjallajökull tephra is about 15 cm thick, while the one produced by the 2011 Grímsvötn event is more than 130 cm.

It is noteworthy that the eruption pattern of Grímsvötn is characterized by distinct 50–80-year-long periods of high eruption frequency alternating with equally long periods of low eruption frequency. The shortest and longest periods are 100 and 160 years and the average is ~140 years. This periodicity coincides with major strain release events on the South Iceland Seismic Zone, and the prediction is that the next 60 years will be an interval of high eruption frequency, peaking around AD 2030–2040.

Grímsvötn volcano also features one of the most powerful geothermal systems in Iceland, with an average output of 4000–5000 megawatts. The geothermal heat melts the ice, and the meltwater accumulates in a 600 m-deep caldera, forming a lake that is sealed by a floating ice shelf. The ice melting gradually increases the volume of the lake, elevating the water table and the surface of the floating ice shelf. When the surface of the lake reaches a critical level, the water pressure is sufficient to open a 50 km-long subglacial waterway extending from Grímsvötn to the edge of Skeiðarárjökull. When this happens, up to 80 per cent of the lake is drained in a jökulhlaup that typically lasts for one to three weeks. Normally the Grímsvötn caldera lake is drained every five to ten years and the jökulhlaups burst onto the sandur plain through the outlet of Skeiðará.

The total volume of the water expelled by these jökulhlaups used to be in the order of 1–3 km³, with a maximum discharge of 5000–40 000 m³/s. However, because of thinning of the ice cap, the volumes of recent outbursts have been much smaller than those that took place before the turn of the century, although every so often, this 'normal' flood pattern is interrupted by volcanically induced jökulhlaups, such as the one generated by the subglacial eruption at Gjálp in 1996. Eruptions such as Gjálp melt large quantities of ice quite quickly and the meltwater is discharged directly into the Grímsvötn caldera. This results in a rapid rise of the water level in the caldera lake and in catastrophic jökulhlaups that burst through the outlets of Skeiðará, Súla and Gýgjukvísl. These jökulhlaups typically cover the whole of Skeiðarársandur and dump large volumes of debris and gigantic blocks of glacier ice onto the sandur plain.

Locality 5.8   Skaftafell national park and Öræfajökull volcano

East of Skeiðarársandur is the Öræfajökull stratovolcano, one of the three giants among Icelandic volcanoes. Towering over the Skaftafell national park, Öræfajökull rises with an average slope of 15° to a summit plateau with an average height of 1850 m. The plateau is an ice-filled summit caldera with an area of 14 km². Several peaks rise above the caldera, including the rhyolite dome Hvannadalshnjúkur, Iceland's highest peak (2110 m). It is the tallest volcano in Iceland and the third largest in terms of volume (285 km³).

Öræfajökull is the most magnificent glacier area in Iceland. From the ice-capped summit, outlet glaciers plunge down the outer slopes as ragged icefalls that have carved steep-sided valleys into the cone. Many of the outlet glaciers extend to the lowland plains at the foot of the mountain, where they have formed spectacular moraines, such as the 100 m-high end moraine, in front of Kvíárjökull. Many of the outlet glaciers feature remarkably regular **ogives** that have earned them a place in the geological history of Iceland. On 11 August 1794, when the naturalist Sveinn Pálsson saw the ogives on Hrútárjökull from the caldera rim, he realized that glaciers had the fluid properties of highly viscous matter and that the outlet glaciers moved down slope by flow.

Most of the rock formations that make up Öræfajökull are younger than 0.7 million years and are typical of the Upper Pleistocene Formation. The volcanic succession is principally made of pillow lavas and hydroclastic

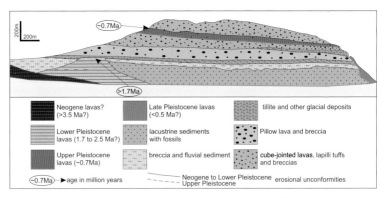

Figure 5.22 Lithostratigraphy of Mt Svínafjall, showing the position of the fossiliferous lacustrine sediments.

tuffs and breccias, along with basalt and andesite lava flows. Rhyolites are present in significant amounts, as both intrusive and extrusive units, and this silicic magmatism appears to have kicked in around 0.2 million years ago. Several of the peaks rising from the caldera rims are remnants of rhyolite lava domes. These volcanic units are intercalated with lesser volumes of tillites, glaciofluvial and glaciolacustrine sediments.

The most spectacular sediment formation within the Öræfajökull succession is the fossiliferous Svínafell interglacial sedimentary beds that outcrop at the base of Mt Svínafellsfjall (Fig. 5.22). The sediment sequence is 120 m thick and consists of well-bedded, fine-grained lacustrine sandstone and siltstone, which contain pollen grains and leaf imprints of alder, birch, rowan, blueberry and crowberry heather, grasses and ferns. The sediments rest unconformably on a 100 m-thick series of basalt-lava flows of Lower Pleistocene age (> 1.7 million years) and are overlain by a thick tillite bed and a series of pillow lavas, subaerial basalt-lava flows and hyaloclastites. The age of the Svínafell sediments falls into the hiatus between basement and the Öræfajökull volcano or in the age range of 0.7–1.7 million years, and the fossilized assemblage indicates vegetation similar to the present-day fauna in Iceland.

Tephrochronology shows that the postglacial volcanic activity at Öræfajökull has been mostly explosive and has not added much to the height and volume of the volcano. Three postglacial flank eruptions produced small scoria cone rows and lava flows on the southern and western slopes. The first historical eruption at Öræfajökull occurred in early June 1362 and was described in a contemporary chronicle as follows:

Volcanic eruptions in three places in the South and kept burning from flitting days [i.e. early June] until the autumn with such monstrous fury as to lay waste the whole of Litla-Hérað, as well as a great deal of Hornafjörður and Lónshverfi districts, causing desolation for a distance of some hundred miles. At the same time there was a glacier burst from Knappafellsjökull into the sea, carrying such quantities of rocks, gravel and mud as to form a sandur plain where there had previously been 30 fathoms [50 m] of water. Two parishes, those of Hof and Rauðilækur, were entirely wiped out. On even ground, one sank in the sand up to the middle of the legs, and winds swept it into such drifts that buildings were almost obliterated. Ash was carried over the northern country to such a degree that foot-prints became visible on it. In addition to this, pumice was seen floating off the west coast in such masses that ships could hardly make their way through it.

Information from other contemporary annals is similar, one of them stating that the floods accompanying the eruption swept away all buildings within the rectory of Rauðilækur except the church. Yet another source states that 'no living creature survived except one old woman and a mare'.

The scanty information from the contemporary chronicles is supplemented by volcanological studies of the eruption products, which show that the vents of the 1362 eruption were within the summit caldera and that the main phase of the eruption was purely explosive. It started with a powerful Plinian phase, which initially sent powerful pyroclastic surges and jökulhlaups cascading down the outer flanks of the volcano. The surges reached beyond 10–15 km from the source vents, travelling at speeds of several hundres metres per second, and wiped out Litla Hérað, a settlement of 35–40 farms situated at the foothills of the volcano at the time (Fig. 5.23a,b). This was followed by tephra fallout from a 30–40 km-high eruption column burying Litla Hérað under decimetres to metres of pumice and dispersing 10 km^3 of rhyolitic tephra towards the east-southeast (Fig. 5.24). This prosperous rural settlement, which may have contained up to 400 inhabitants, was laid waste by the eruption and remained abandoned for decades. When a revival came at last, the district assumed a new name. It was called Öræfi, the name it still carries today, which literally means

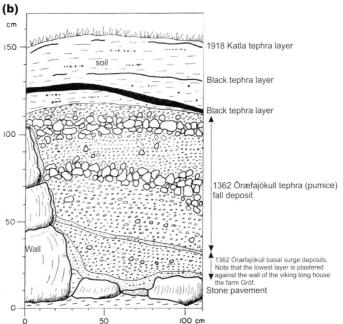

Figure 5.23 (a) Photograph showing the internal stratigraphy of the 1362 Öræfajökull tephra deposit adjacent to the farm ruins at Bær, which was more than 12 km south of the source vents. The lowest part of the tephra is comprised of deposits laid down by fast-moving pyroclastic surges. (b) A drawing by the late Professor Sigurður Þórarinsson of the 1362 Öræfajökull tephra deposit at the ruins of the farm Gröf, which was at the western foothills of the volcano and more than 9 km from the source vents.

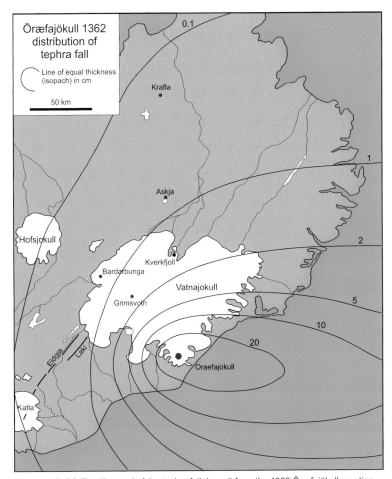

Figure 5.24 The dispersal of the tephra fall deposit from the 1362 Öræfajökull eruption.

'wasteland'. The regular and unidirectional distribution of the tephra fall indicates that the initial phase was brief, but powerful enough to cause tephra fall on mainland Europe and in Scandinavia. The 1362 Öræfajökull is the largest explosive rhyolite eruption in Iceland in historical times, and its magnitude was similar to that of the 1991 eruption of Mt Pinatubo in the Philippines. The second and most recent historical eruption of Öræfajökull occurred on 3 August 1727 and lasted until April or May 1728.

Locality 5.9
Breiðamerkursandur and Jökulsárlón glacial lagoon

Breiðamerkursandur [64.0391, −16.2681] is the collective name for the outwash plains in front of *Hrútárjökull* [64.0071, −16.4555], *Fjallsárjökull* [64.0163, −16.4056] and *Breiðamerkurjökull* [64.1086, −16.2988]. These glaciers reached their maximum postglacial extent during the latter half of the nineteenth century, forming a classic piedmont glacier. In 1894 the shortest distance between the edge of Breiðamerkurjökull and the coastline was only 256 m. At the *River Jökulsá* [64.0437, −16.1794], the distance between the glacier and the sea was about 0.5 km. Since the mid-1890s these glaciers have been retreating. As of 2018 Breiðamerkurjökull has retreated 7.25 km (equal to about 60 m/year) and in doing so, has formed the now popular tourist destination, Jökulsárlón. Between 1894 and 1968 the front of the piedmont glacier retreated about 2 km, breaking it up into the three outlet glaciers we see today. At the time of settlement in Iceland (874–950), the front of Breiðamerkurjökull was at least 10 km behind the 1894 end moraines. At that time, two farms were established on the western side of Breiðamerkursandur. One of them, named Fjall, was located at the foot of *Fellsfjall* [64.1294, −16.1367], and the other, Breiðá (with a church), was located a little farther to the east. Both farms were destroyed between 1695 and 1720 by the advance of the glaciers during a cold snap of the Little Ice Age. The retreat of the Breiðamerkurjökull since the 1890s has produced the glacial lagoon, *Jökulsárlón* [64.0537, −16.1806], which reaches a depth of about 110 m below sea level. Furthermore, when crossing the sandur plains it is noteworthy that all glacier rivers on Skeiðarársandur and Breiðamerkursandur have begun to erode the upper parts of the plains, and upper segments of the river channels have stabilized. The braided part of the channel systems is now only present in the lower reaches of the sandur plains close to the coast. The cause of this change can be linked to the retreat of the outlet glaciers and formation of lagoons in front of their snouts. The net effect is that the sediment-laden meltwater discharge from the glaciers now dump the bulk of their sediment load into the lagoons, thus increasing the erosional power of the outlet water.

Chapter 6

The southeast and east

General overview

There is a striking change of scenery to the east of Breiðamerkurjökull as we pass into the older Neogene Basalt Formation (Figs 6.1 and 6.2). Here the main outlet glaciers of the Quaternary icecap have honed and enlarged pre-existing fluvial ravines to carve deep U-shaped valleys into the gently dipping basalt-lava plateau, thereby producing a classic alpine landscape (Fig. 6.3). Many of these glacially carved valleys are so deep that they reach well below sea level and form a fjord landscape, which is the most distinctive feature of East Iceland. Typically, the main outlet glaciers were fed by many smaller tributary glaciers. These tributary glaciers carved out the hanging valleys and **corries** (cirques) that line the mountain slopes above the valleys and fjords (Fig. 6.4). Good examples are found in Berufjörður, Stöðvarfjörður and Fáskrúðsfjörður, where in some instances the tributary glaciers have cut so far back into the mountainside that the corries of these adjacent fjords are separated by only an extremely thin serrated ridge known as an arête.

Eastern Iceland provides a continuous 8 500 m-thick succession through the Neogene Basalt Formation. This succession covers about 13 million years of geological history (from about 16 to 3 million years ago; Fig. 6.2), indicating an average accumulation rate of about 650 m per million years. The Neogene Basalt Formation was constructed in much the same way as the volcanic successions in the active volcanic zones and it comprises the same geological elements, with the exception of subglacial volcanic and sedimentary rocks. The volcanic system is the principal building block of the succession and is expressed in the succession as 10–30 km-wide and 50–100 km-long lenticular stratigraphical units. So far, a total of 17 extinct volcanic systems have been identified, each made up of an extinct central

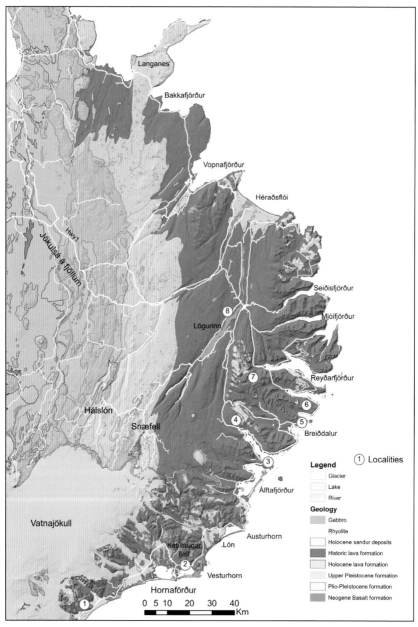

Figure 6.1 The main geological features of East Iceland. Also shown are excursion routes, stops and sites.

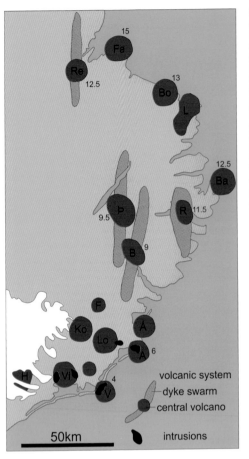

Figure 6.2 Occurrence and distribution of extinct volcanic systems in East Iceland. The central volcanoes are abbreviated as follows: Fa, Fagridalur; Re, Refstaðir; Bo, Borgarfjörður eystri; L, Loðmundarfjörður; Ba, Barðsnes; R, Reyðarfjörður; Þ, Þingmúli; B, Breiðdalur; Á, Álftafjörður; A, Austurhorn; Lo, Lón; Ko, Kollumúli; F, Flugustaðardalur; V, Viðborðsfjall; H, Hoffellsjökull. The numbers next to each volcanic complex indicate their approximate age in millions of years.

volcanic system

dyke swarm

central volcano

50km

intrusions

Figure 6.3 Kálfafellsdalur is a U-shaped valley carved out by an Ice Age valley glacier where the Miðvötn–Steinavötn River now runs down the bottom of the valley, flanked by 1000–1500 m-high mountain slopes. Brókárjökull (centre) is a small outlet glacier from Vatnajökull icecap which cascades down the cliffs at the bottom of the valley.

Figure 6.4 Northern slope of Búlandstindur in Berufjörður featuring a well-formed corrie (cr) with an arête (ar) above it. The white arrows point towards the layer-cake appearance of the basalt lava flow stratigraphy.

volcano and/or a dyke swarm. The dyke swarm is the subsurface representation of the fissure swarms within the currently active volcanic zones. A majority of the Neogene Formation is made up of fairly regularly stacked and relatively thin (3–20 m-thick) and laterally extensive basaltic lava flows, giving the formation a layer-cake appearance (e.g. Fig. 6.4). However, the successions that make up the central volcanoes comprise abundant (> 20%) intermediate and rhyolite volcanic units, which typically are bulkier (tens to hundreds of metres thick) and more irregular in their form. These units disrupt the apparent layer-cake stratigraphy of the basaltic lava. Therefore, these more evolved central volcano formations are easily distinguished from the sequences dominated by basalt lava that were produced by dyke-fed fissure eruptions.

We present two excursions in Southeast and East Iceland. The first one takes us across Southeast Iceland from Breiðamerkurjökull to Álftafjörður. The second excursion deals with the Neogene Basalt Formation in East Iceland from Berufjörður to Fljótdalshérað.

Suðursveit–Lón–Álftafjörður

This excursion follows Highway 1 from *Breiðamerkurjökull* [64.0546, −16.4606] to *Álftafjörður* [64.5751, −14.5418] and includes the districts of Suðursveit, Mýrar, Hornafjörður and Lón.

Locality 6.1
Suðursveit and Mýrar – valley glaciers and alpine landscape

In Suðursveit and Mýrar the cultivated lowlands rise only a few metres above sea level. To the south, they are separated from the ocean by a series of coastal lagoons and to the north the steep and rugged mountains shelter them from the Vatnajökull icecap. Despite their lush green, the narrow stretches of farmland were originally sandur plains, built from debris transported by glacial rivers such as *Steinavötn* [64.1654, −15.9765], *Kolgríma* [64.2462, −15.6743] and *Hornarfjarðarfljót* [64.3550, −15.3669].

The area is a classic example of alpine landscape, where the mountains are incised by a series of steep-sided valleys carved out by descending outlet or valley glaciers. The glaciers have disappeared from some valleys, whereas others are filled to the brim by outlet glaciers from the Vatnajökull icecap. The 10 km-long *Kálfafellsdalur* [64.1868, −15.9488] is a spectacular example of these valleys, where the Miðvötn–Steinavötn River runs down the narrow sandur plain flanked by 1000–1500 m-high mountains, which are among the tallest in Iceland. At the bottom of the valley, a small glacier, *Brókárjökull* [64.2577, −16.1390] (Fig. 6.3), cascades off an 800 m cliff face to the valley floor, forming a stunning icefall. Occasionally, large ice blocks tumble down the steep face of the glacier, accompanied by a rumble that reverberates throughout the valley. The surrounding mountains are composed almost entirely of basalt flows, which originally formed an extensive lava plateau that later was carved out by the Ice Age glaciers. These flows are some of the youngest lavas of the Neogene Basalt Formation and they formed by repeated effusive eruptions at three to five million years ago. Deep in the valley many inclined dykes cut through a series of rhyolite and basalt lavas, indicating the presence of an extinct central volcano.

Farther east, the large valley glaciers Skálafellsjökull, Heinabergsjökull and Fláajökull stretch their snouts onto the coastal sandur plain and in doing so demonstrate what the situation was like at Kálfafellsdalur at the time of its formation.

On the way

Mt *Borgarhöfn* [64.1962, −15.7805] (or Hestgerðismúli), the promontory to the east of Kálfafellsdalur, is crowned by a magnificent colonnade. These are the columnar joints of a 15–20 m-thick basalt-lava flow that is part of the so-called Múli Formation. This formation belongs to the Plio-Pleistocene

Múli Formation

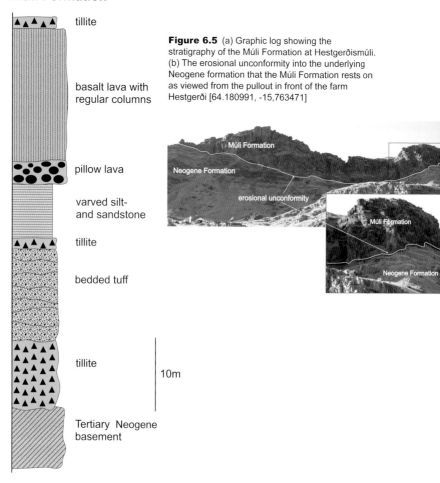

tillite

basalt lava with regular columns

pillow lava

varved silt- and sandstone

tillite

bedded tuff

tillite

10m

Tertiary Neogene basement

Figure 6.5 (a) Graphic log showing the stratigraphy of the Múli Formation at Hestgerðismúli. (b) The erosional unconformity into the underlying Neogene formation that the Múli Formation rests on as viewed from the pullout in front of the farm Hestgerði [64.180991, -15,763471]

Formation (< 3 million years old) and comprises the youngest rocks in this region (Fig. 6.5). The Múli Formation rests on an eroded Neogene basement, which is made up of a sequence of 4–20 m thick lava flows commonly separated by decimetre-thick red to brown sedimentary interbeds. The Múli Formation consists of the above-mentioned columnar lava, which grades downwards into pillow lava and **kubbaberg** breccia. These volcanic rocks are capped by a thin tillite bed and are underlain by well-bedded lake sediments. Below the lake sediments is a thin diamictite bed and then a unit of layered móberg tuff. At the base is a thick tillite unit that rests directly on the eroded Neogene basement. The Múli Formation was formed in a valley eroded into the Neogene basement, and the three tillite beds at the foot and top of the sequence indicate that on at least three, and most likely more, occasions the valley was congested by glaciers. The complete sequence also shows that several volcanic eruptions contributed to the construction of the Múli Formation. The stratified móberg tuff is an example of a subglacial eruption that occurred when the first glacier occupied the valley. The lava flow rests directly on lake sediments, which suggests that the lava flowed into a small lake located in front of a valley glacier. The pillow lava and kubbaberg breccia at the base of the lava were formed when it initially came into contact with the lake water, and their thickness shows that the water depth was not more than 2 m. When this pedestal had been formed, the rest of the lava advanced forwards on dry land and, as the lava cooled and solidified, formed the spectacular columnar joints. Finally, the glacier rushed forwards once again, forming the topmost tillite bed. Rocks of corresponding age are re-emerging, but further north along the margins of the retreating Vatnajökull glacier (Fig. 6.1) and may reveal a more complete story on the formation of the Múli Formation, and perhaps provide clues about the origin of the off-rift volcanism that produced it.

Locality 6.2   Hornafjörður to Lón – a trip through a volcano

Traversing the region from *Hornafjörður* [64.2534, −15.2089] across *Lón* [64.4085, −14.6620] to *Álftafjörður* [64.5795, −14.5489] is probably as close as one can get to travelling through a volcano. It is difficult to imagine that the peaceful countryside of Lón was once the boiling hot roots of three, now extinct, volcanoes which, at the time, were located at depths of 2–3 km below the surface.

Here, erosion has removed much of the extrusive rock and exposed deeper intrusions. The larger gabbro intrusions, such as those exposed at *Viðborðsfjall* (on the northeast side and not visible from Highway 1), [64.3437, −15.4171], *Ketillaugarfjal* [64.3422, -15.2271], *Vesturhorn* [64.2767, −14.9548] and *Austurhorn* [64.4122, −14.5439] (Fig. 6.1), represent ancient basaltic magma chambers that provided the fuel to some of the Neogene central volcanoes in this region when they were active about four to six million years ago (Fig. 6.6). Eruptions from these magma chambers produced many of the dykes that are seen in the area, including the one that is visible face-on on the south side of Viðborðsfjall, and which formed basaltic lava sequences. Several granophyre intrusions (e.g. at

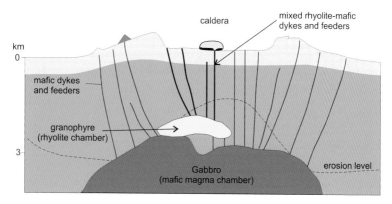

Figure 6.6 Simplified cross section of a Tertiary central volcano and its roots, showing the relative position of basaltic (gabbro) and rhyolite (granophyre) magma-holding chambers and distribution of dyke intrusions.

Ketillaugarfjal, *Slaufrudalur* [64.3238, −15.0081], Vesturhorn and Austurhorn) are also exposed here. Typically, they are situated a little higher in the stratigraphy and draping the gabbro intrusions. These granophyre intrusions represent smaller holding chambers that at the time contained rhyolite magma formed by partial melting of the surrounding basaltic crust. Eruptions from these chambers were responsible for formation of the rhyolite lavas and pyroclastic flows.

The Austurhorn intrusion is a composite gabbro and granophyre intrusion that outcrops over 11 km². It represents one of the solidified magma chambers in the Lón region that was active about six million years ago (Fig. 6.2). The eastern part of this intrusion consists of a net-veined complex, formed by intimate mixing of mafic and acid magma (Fig. 6.7a). In the net-veined complex the mafic magma occurs as pillow-like or angular masses enclosed in and veined by silicic magma (granophyre; Fig. 6.5b). This complex was formed as the hotter mafic magma intruded into semi-molten granophyre magma situated near the chamber roof. The fine-grained margins of the pillow-like bodies show that the mafic magma was quenched against the colder granophyre magma.

On the way

Travelling onward from Austurhorn we enter the domain of the Álftafjörður volcano (Fig. 6.2), which features the spectacular *Mælifell* [64.4722, −14.5108] caldera, a subsidiary structure on the southern margin of the main caldera of the Álftafjörður volcano. The Mælifell caldera is 2 km in diameter

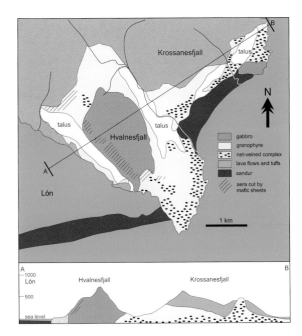

Figure 6.7 (a) Geology of the Austurhorn gabbro intrusion: rock types and their outcrop pattern. (b) Close-up view of the net-veined complex showing mafic (dark) inclusions in a granophyre host. In the centre of the photo two pillow-like mafic inclusions are cut by a later silicic vein.

Basalt inclusion (dark) with convoluted (fluidal) outlines, which indicating molten state upon inclusion in the felsic magma (light) surrounding it (now solidified). Later when tis body had solidified, it was cut by an inclined felsic sheet (light).

and is made up of inward-dipping welded tuffs, agglomerates and tuffaceous sediments, along with rhyolite and andesite lava flows. The welded tuffs occur as thin sheets of pitchstone (obsidian-like rock) and as much thicker units of rhyolite lava-like masses. The pitchstone sheets contain basalt-glass (melt) inclusions, indicating concurrent eruption of rhyolite and basalt magmas. These relationships, along with those observed at Austurhorn (see above), imply that the rocks at Mælifell were formed in eruptions triggered by intrusion of hot mafic melt into a cooler felsic magma body.

The eastern fjords

This excursion takes us across the Neogene Basalt Formation in East Iceland from *Berufjörður* [64.7599, −14.4369] to *Fljótdalshérað* [65.2256, −14.5329].

Locality 6.3   The Neogene dyke swarms

The Neogene succession is dissected by many subparallel basalt dykes, such as those cutting through the cliffs above the *Kárastaðir* [64.6972, −14.2270] farm on the north side of Berufjörður (Fig. 6.8). Here the near-vertical dykes are 3–20 m thick and they display the horizontal jointing arrangement that is a characteristic feature of dyke intrusions. The joints form because the magma shrinks upon cooling and solidification; they are always perpendicular to the cooling surface. When the magma intrudes the

Figure 6.8 Dykes (vertical structures) cutting through the basalt lava succession in Berufjörður, East Iceland.

bedrock as a vertical dyke, the joints have horizontal orientation because the principal cooling surfaces are vertical.

The dykes in the Neogene Basalt Formation typically occur in swarms 5–15 km wide and over 50 km long, and represent the roots of once-active fissure swarms like those seen on the surface within active rift zones. The

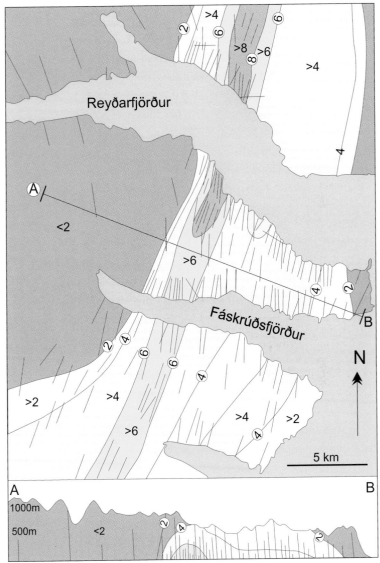

Figure 6.9 Map showing variation in intrusion densities across Reyðarfjörður dyke swarm, East Iceland.

dyke swarms are typically associated with central volcanoes, such as the Breiðdalur volcano described below (Fig. 6.2). Each swarm consists of hundreds of dykes. The number density is highest along the central axis of the swarm and decreases laterally to the margins (Fig. 6.9). It also increases with depth below the original surface, because only about 25 per cent of the dykes reach the surface and become feeders to eruptions; the remainder are arrested within the crust, where they cool and solidify.

By measuring the collective thickness of dykes along a horizontal line over a given distance (e.g. 1000 m), one can calculate how volume is occupied by the dykes. Such measurements in East Iceland show that the dykes most commonly make up 5–6 per cent, meaning that the cumulative thickness of the dykes amounts on average to 50–60 m per 1000 m of crust. In the centre of the swarms, they can occupy up to 28 per cent of the crust. The significance of these measurements is revealed when we think of the plate movements. The space occupied by the dykes is a direct measure of how much spreading occurred because of crustal rifting when these systems were active.

Locality 6.4   Breiðdalur central volcano

The *Breiðdalur* [64.8121, −14.1912] volcano is a classic example of an extinct and eroded volcanic centre (Fig. 6.10). It has a volume of 400 km^3 of mafic, intermediate and felsic lavas and pyroclastics that have a maximum thickness of about 2000 m. The central vent area of the Breiðdalur volcano is manifested as a profusion of rhyolite and andesite lavas and pyroclastic rocks, which are flanked by thin mafic lava flows dipping (5–10°) away from the central vent area. At the time of activity, the Breiðdalur volcano was a 500–1000 m-high cone rising above the surrounding lava plain, which was constructed by eruptions of the associated fissure swarm. The *Röndólfur* [64.8023, −14.3702] Group is a series of conspicuous peaks (i.e. Stöng, Slöttur, Röndólfur, Smátindur and Flögutindur) on the eastern flank of the volcano, made up of rhyolite lavas and feeder dykes.

Locality 6.5   The Lambafell sill

The mountain that stands tall over the town of *Breiðdalsvík* [64.7761, −14.0302] to the north features the peaks Snæhvamstindur, Lambafell and Súlur. An impressive mafic (dolerite) sill can be seen cross-cutting the layered basaltic lava flow sequence near the top of the mountain (Fig. 6.11). From this viewing angle it is easy to imagine how this ~100 m-thick magma

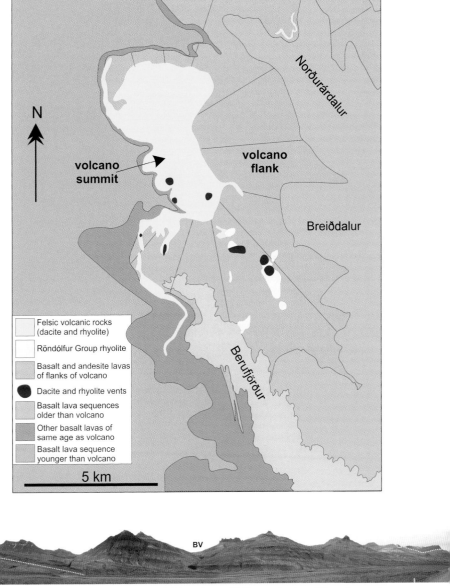

Figure 6.10 (a) Geology of the Breiðdalur volcano. (b) Panoramic view of the outcrop cut through the Breiðdalur volcano (BV). Broken lines indicate the approximate bottom (far left) and top (far right) of the volcano. View is to the south and photo taken in Suðurdalur.

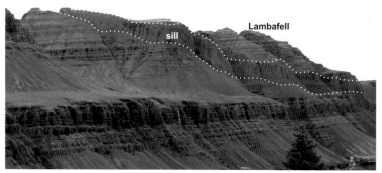

Figure 6.11 Lambafell sill (s) cross-cutting the lava sequence above the town of Breiðdalsvík. View is to the NE. The white dotted lines outline the sill margins.

body forced its way in a stepwise fashion up through the lava sequence around 10 million years ago, migrating ever closer to the surface, which at the time was not much more than 700 m above the present mountain top.

Locality 6.6   The Sandfell laccolith

Along the south side of Fáskrúðsfjörður, *Mt Sandfell* [64.8765, −13.8972] attracts immediate attention because of its pale rocks, in stark contrast with the dark basalt lavas that make up the bulk of the stratigraphical succession here. The pale rock at Sandfell is rhyolite. The steeply dipping strata adjacent to the rhyolite body show that it intruded the basalt-lava series about 11.7 million years ago and it is a stunning example of shallow intrusions known as **laccoliths** (Fig. 6.12). At the time, the slowly rising rhyolite

Figure 6.12 Sandfell (S) laccolith. Steeply dipping strata on either side of Sandfell were pushed up by the intrusion.

magma came within 500 m of the surface before it began to accumulate as an intrusion. The magma made space for itself by lifting up the overlying strata, forming a dome-shaped hill that in the end rose 400–500 m above its surroundings. A small amount of magma made it to the surface through cracks in the up-domed roof. This type of intrusion is also known as a cryptodome, because of its resemblance to rhyolite lava domes, which, however, are formed by extrusions onto the surface.

Locality 6.7   The Skessa tuff and other Neogene ignimbrites

The Skessa rhyolite tuff is the largest known ignimbrite in Iceland (Fig. 6.13). It can be found in outcrop over an area of 430 km² and has a volume of 4 km³, which exceeds the dense rock volume of the rhyolite tephra produced by each of the Plinian eruptions at Hekla 3100 and 4200 years ago. It was formed by a pyroclastic flow that originated from an explosive Plinian eruption at vents possibly located below Mt Röndólfur in the Breiðdalur

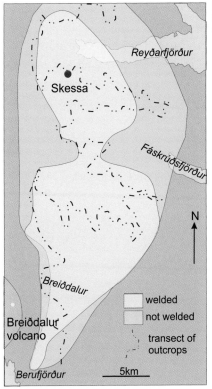

Figure 6.13 The Skessa tuff. A map showing the dispersal of the ignimbrite sheet.

region. Indeed, it has been suggested that this eruption may have initiated the growth of the spectacular Breiðdalur central volcano (see p. 164). The Skessa tuff is thickest closest to its source vents (about 10–12 m) and gradually tapers off to less than 2 m at its distal end just north of Reyðarfjörður. Thus, the pyroclastic flow swept northwards across the lava plains and the hot murky cloud overwhelmed about 500 km² of country.

At its type locality, about 500 m up the north face of *Mt Skessa* [65.0082, −14.2760] in Reyðarfjörður (Fig. 6.13), it is the only felsic rock exposed in a 1000 m-thick section of basalt lavas. Here the Skessa tuff occurs as a greenish layer about 3 m thick, sandwiched between two basalt-lava flows. However, here and in outcrops closer to the source vents, the ignimbrite exhibits a distinctive internal layering, as shown in Figure 6.14, whereas farther to the north this layering is absent. A more accessible, but partial outcrop into the Skessa tuff is at *Skriðuá* [64.8104, −14.3335] in Breiðdalur.

The internal layering results from welding of parts of the ignimbrite during and immediately after its emplacement. In other words, upon deposition, the interior of the ignimbrite was still hot enough for the tephra fragments to be fused together and cause the highly vesicular pumices to collapse and flatten and thus form the flame-like **fiammés** that give the rock a distinct flow-like appearance. This process also hardens the rock and gives the welded part a lava-like appearance, but the **eutaxitic texture** and presence of fiammé reveal its true origin. In general, the thickness of the welded horizon increases to the south towards the source vents. It also shows that, when the ignimbrite was emplaced at Skessa some 30 km north of the source vents, it was still very hot. The general distribution of different rock types in the Skessa tuff with distance from source is depicted in Figure 6.14.

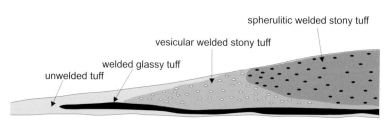

Figure 6.14 The Skessa tuff. The internal structure of the ignimbrite. At the base is a veneer of non-welded tuff that is overlain by thin, pale grey, glassy welded tuff. The grey welded tuff grades sharply upwards into thicker pinkish stony welded tuff, which is capped by a thin layer of non-welded tuff.

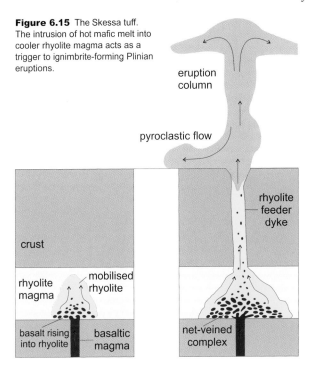

Figure 6.15 The Skessa tuff. The intrusion of hot mafic melt into cooler rhyolite magma acts as a trigger to ignimbrite-forming Plinian eruptions.

In some places (e.g. *Tinnudalur* [64.8717, −14.1817], Fig. 6.13), tree moulds are preserved in the non-welded base of the ignimbrite. This shows that the pyroclastic flow decimated everything in its path, and this eruption must have had a devastating impact on the vegetation at the time.

About 2 per cent of the fragments in the Skessa tuff are vesicular basalt clasts, showing that a little mafic magma was erupted with the rhyolite. This suggests that sudden heating of rhyolite magma at depth by much hotter mafic magma triggered the Plinian eruption that produced the Skessa ignimbrite (Fig. 6.15). Ignimbrite sheets are fairly common in the Neogene Basalt Formation; they are very similar in their make-up to the Skessa tuff and probably formed by a similar mechanism. In East Iceland the rhyolitic rocks make up 10–12 per cent of the stratigraphical succession, and the ignimbrites represent about a third of that volume.

Locality 6.8   Lögurinn

In the district of Fljótdalshérað, *Lögurinn* [65.1674, −14.651] ('the loch') is East Iceland's largest lake and the third largest in Iceland as a whole. It is about 25 km long, 3 km^2 in area, with a maximum depth of 112 m. The lake is in a U-shaped glacial valley that extends more than 90 m below present-day sea level. Tales from the Middle Ages tell a story of a monster that supposedly lives in Lögurinn, similar to the one in Loch Ness in Scotland. Modern-day glimpses are rare and the sighting of the monster at the surface appears to be directly related to events when methane gas escapes from the bottom sediments and bubbles up through the water column.

Chapter 7

The northeast

General overview

The main feature of Northeast Iceland is the North Volcanic Zone and the Tjörnes Fracture Zone that links it with the offshore Kolbeinsey Ridge (Fig. 7.1). There are five active volcanic systems within the 70 km-wide and 200 km-long North Volcanic Zone, including the infamous Krafla and Askja systems (see Fig. 1.5). The Tjörnes Fracture Zone is a 70 km-wide and 120 km-long belt of faults that connects the North Volcanic Zone with the southern tip of the submarine Kolbeinsey Ridge (see Fig. 1.2). It is an active fracture zone, and faulting associated with displacement of the crust is capable of producing earthquakes of magnitude 6–7 on the Richter scale.

We present two excursions in Northeast Iceland. The first visits the canyon country of Jökulsá á Fjöllum and the remarkable sediment succession at Tjörnes. The second excursion is centred on the Lake Mývatn area – one of the most impressive natural wonders in Iceland.

Jökulsá–Tjörnes

This excursion begins at the junction of Highway 1 and Road 864, and follows the canyon of Jökulsá á Fjöllum north to Ásbyrgi. Then it takes us along Road 85 across Axarfjörður to the Tjörnes Peninsula and the town of Húsavík.

Locality 7.1   Jökulsárgljúfur, Dettifoss and Ásbyrgi

Jökulsárgljúfur [65.8213, −16.3848] is the canyon of the River Jökulsá á Fjöllum and it stretches for 30 km from the Selfoss waterfall to the place where the river emerges from the gorge at the head of the sandur plain at the bottom of Öxarfjörður. It offers some of the most magnificent scenery in Iceland. Lavish vegetation thrives in sheltered spots among the regular jointed basalt-lava flows that make up the canyon walls, which sometimes

resemble a giant stone-pillar fence erected by forces of another world
(Fig. 7.2a). The crystal-clear spring water that issues from cracks in the wall
is in stark contrast to the brownish glacial water rushing forth at the bottom
of the canyon.

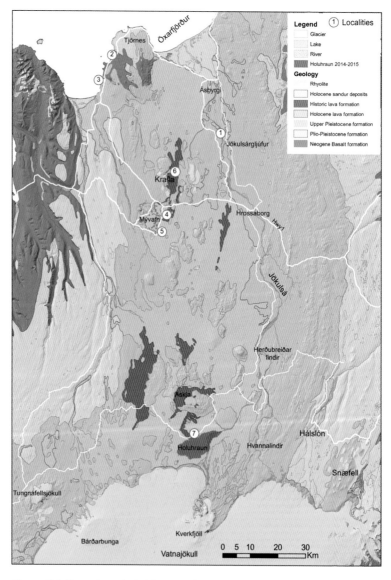

Figure 7.1 The main geological features of Northeast Iceland including the new 2014–15 lava flow field at Holuhraun. Also shown are excursion routes and sites.

Strictly speaking, the Jökulsárgljúfur is not entirely a canyon. It can be divided into three main parts. The southernmost part, 9 km-long, is a clear canyon with near-vertical walls and carved out by fluvial erosion that follows fault lines trending north–south. Here the canyon walls comprise six lava flows and two sedimentary units (Fig. 7.2b). The upper sedimentary unit occurs between lavas 3 and 4. Undercutting and collapse of lava 3 caused by erosion of this sedimentary unit resulted in the formation the Selfoss waterfall. The second sedimentary unit occurs at the foot, in the deepest part of the canyon where two very thick lava flows (44 m in total)

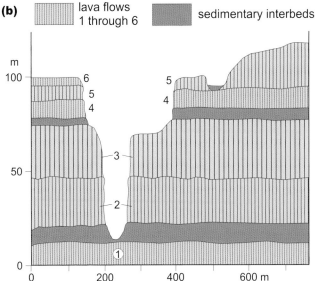

Figure 7.2 (a) Jökulsárgljúfur (Jökulsá canyon). View is to south. (b) Geological cross section of Jökulsárgljúfur.

overlie it. Undercutting and collapse of these flows, induced by erosion of this lower unit, have formed *Dettifoss* [65.8141, −16.3846], the greatest waterfall in Iceland in terms of volume discharge. The middle part of Jökulsárgljúfur is a 9 km-long U-shaped valley with boulder and gravel terraces at different heights on both sides. The northernmost part of Jökulsárgljúfur is again a real canyon, 11.5 km long.

About 2 km to the northwest of the mouth of Jökulsárgljúfur is *Ásbyrgi* [66.0090, −16.5060], a stunning impression into the land surface, which, according to the Viking mythology, was created when Sleipnir (Óðinn's horse) accidentally put one of his eight legs down on the ground. Ásbyrgi is an oval depression bounded by up to 100 m-high vertical walls on three sides, with an opening facing north. In its centre rises a cliff-bounded island.

Jökulsárgljúfur, including Ásbyrgi, appears to have been formed in three main periods of intense canyon cutting, one in the period 9–10 thousand years ago, another around 5 thousand years ago, and the last one at about 2000 years before present. The floods, 16 in total, originated beneath the Vatnajökull icecap in the region in and around Kverkfjöll volcano (at least 6 floods) in the east to the Bárðabunga volcano in the west, and were probably triggered by a subglacial eruption. Recent studies indicate a peak discharge of about 40 000 m^3/s, when transporting water from the source to the sea at Öxarfjörður, a distance of 170 km.

The *Sveinar* [65.6858, −16.4588] graben system runs almost directly north–south for about 30 km and crosses the Jökulsá Canyon 4 km north of Dettifoss waterfall. West of the Jökulsá River, the cone row (volcanic fissure) *Rauðuborgir* [65.6012, −16.4743] runs along the centre of the graben system. On the east side of the canyon the graben system continues in the same direction, whereas the *Randhólar* [65.8491, −16.3780] cone row, which is a direct continuation of the volcanic fissure on the western side, sways to the northeast away from the graben. It is likely that the Sveinar graben was already in existence when the Rauðuborgir–Sjónpípa–Randhólar fissure eruption took place in the early Holocene. It appears that the eruption started to the east of the river, and the fissure propagated to the south-southwest until it was captured by the graben and then continued within it towards the south. Where the river crosses the Rauðuborgir–Randhólar cone row, just north of the Dettifoss waterfall, the canyon cuts through one of the cones of the fissures, exposing its interior and the top 100 m of the feeder dyke (Fig. 7.3). This outcrop provides a

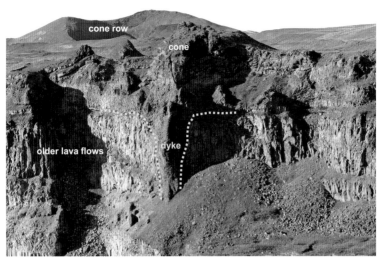

Figure 7.3 Section through the Rauðuborgir cone and its feeder dyke in the eastern wall of Jökulsárgljúfur.

unique picture of a cone row in three dimensions. At *Hljóðaklettar* [65.9464, −16.5330], about 15 km north of Dettifoss, one of the lava flows produced by the Rauðuborgir–Randhólar fissures advanced onto wetlands, and in doing so initiated rootless eruptions. Hljóðaklettar and the surrounding lava crags are the lava remnants of the rootless cone group that at one time decorated this part of the land.

Locality 7.2   Tjörnes sediment succession – transition from Pliocene to Pleistocene climate

At *Tjörnes* [66.1519, −17.0866] is a 1200 m-thick near-continuous stratigraphical section that records about three million years of Iceland's geological history, extending from the late Pliocene (at four million years) and into the Upper Pleistocene (0.7 million years). This remarkable succession provides an extensive record of how the fauna, flora and climate changed throughout Plio-Pleistocene times. The succession at Tjörnes features thick sequences of fluvial, estuarine and fossiliferous marine deposits, which are intercalated with several series of lava flows and some tillite beds. It is divided into three main sequences, which from oldest to youngest are the Tjörnes (proper), *Furuvík* [66.1762, −17.2451] and *Breiðavík* [66.1867, −17.1563] groups (Fig. 7.4).

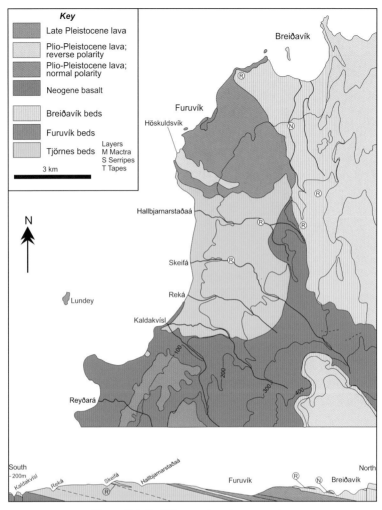

Figure 7.4 The Tjörnes sedimentary succession.

The Tjörnes Group rests on an eroded lava pile of the Neogene Basalt Formation, formed more than eight million years ago, as can be observed in outcrop in the Kaldakvísl ravine on the western shores of the peninsula. The cliff sections along the coast offer excellent exposure through the group, where it can be followed in a continuous 6 km-long outcrop north to Höskuldsvík (Figs 7.4, 7.5a). The beds dip 5–10° to the northwest and are dislocated by faults in a few places. The cumulative stratigraphical thickness of the Tjörnes Group is about 500 m and it consists mainly of marine

Figure 7.5 The Tjörnes sedimentary beds: (a) view of the Tapes layers in coastal cliffs at Tjörnes; (b) sedimentary bed rich in fossil shells.

sediments rich in fossilized shells and snails (Fig. 7.5b). Intercalated with the marine beds are fluvial and lacustrine sediments, as well as occasional lignite seams.

The oldest sediments of the Tjörnes Group are about four million years old and the youngest about 2.5 million years old. It is divided into three sedimentary sequences on the basis of key reference fossils (i.e. mollusc species that distinguish each sequence). The oldest are the *Tapes* layers, then the *Mactra* layers, followed by the *Serripes* layers (Fig. 7.4).

Tapes

The *Tapes* layers, which outcrop between the *Kaldakvísl* [66.1017, 17.2757] and *Reká* [66.1146, 17.2504] streams, comprise a sediment sequence consisting of alternating shell-rich marine and lignite beds. Both are formed in a nearshore environment, the former in shallow coastal water and the latter above sea level in marshy areas adjacent to the shore. Today, the index animal, *Tapes*, lives no farther north than the North Sea, indicating much warmer seas around Iceland at the time of the sediments' formation (i.e. 4–3.5 million years ago) than today.

Mactra

The *Mactra* layers that are found in the area between the streams Reká and Hallbjarnarstaðará take their name from an extinct mollusc species. This sequence is also characterized by alternations of shell-rich deposits and lignite seams, and the sediments were deposited in an environment similar to that of the *Tapes* layers. Some of the species found in the *Mactra* layers live today in much warmer seas, mainly in the region between the North Sea and the Canary Islands. Fossil pollen from the lignite seams shows that coniferous forest was growing on land surrounding the marshy areas. In addition to spruce, pine and larch, broadleaf trees such as oak and beech were growing in these forests, along with alder, birch and willow. Thus, both the fossil marine fauna and the terrestrial flora indicate a much warmer climate than at present. The winters were mild, with mean monthly temperatures above 0°C, and the waters around Iceland were at least 5°C warmer than today.

Serripes

The *Serripes* layers, which outcrop in the coastal banks between *Hallbjarnarstaðará* [66.1416, −17.2424] and *Höskuldsvík* [66.1569, −17.2780], account for just over half of the thickness of the Tjörnes beds. The sequence consists almost entirely of marine sediments, with a few thin lignite seams towards the top. The lava that caps the *Serripes* sequence is 2.5 million years old. The fossil fauna at the base of the *Serripes* layers shows that most of the warm-water molluscs have disappeared from the waters around Iceland and are replaced by molluscs that prefer much colder seas. However, the sea temperature must have been somewhat higher than at present, because some warm-water species (e.g. *Solenensis*) were still living off the north coast of Iceland at that time. The most astonishing thing about this change in the fauna is that, out of 100 fossilized species of mollusc, about 25 per cent originated in the Pacific Ocean. These molluscs migrated from the Pacific across the Arctic into the Atlantic Ocean when the sea flooded the Bering Strait for the first time. Most of the Pacific molluscs are cold-water species, including *Macoma* and *Serripes grönlandicus*, which is the index fossil for this sedimentary sequence. Therefore, the sudden cold-climate appearance of the fauna at the base of the *Serripes* layers may be a false alarm, because it probably does not indicate onset of sudden climatic cooling. It is more probably an artefact of the path the molluscs followed as they migrated from Pacific to Atlantic. The only molluscs that made it across were those that could survive in the colder Arctic seas. On the other hand, the opening of the Bering Strait would have caused some changes in the ocean-current circulation and thus the climate. This change may have enhanced the cooling trend that was already in progress, because, shortly afterwards, the frigid winter of the Ice Age arrived in full force.

The Tjörnes and Breiðavík groups are separated by a 250 m-thick series of basalt-lava flows that represents about half a million years of volcanic activity (2.5–2 million years ago). The Furuvík Group is in the middle of this lava series (Fig. 7.4) and is composed of two tillite beds with a cumulative thickness of about 40 m. These are the oldest glacial remains at Tjörnes and they show that the regional ice sheet extended to the northern shores of Iceland about 2.2 million years ago. This is taken to mark the onset of full glaciation in Iceland (see Fig. 1.21).

The Breiðavík Group rests on the lavas that cover the Furuvík deposits (Fig. 7.4) and comprises a 150 m-thick sequence of fossiliferous marine sediments alternating with tillite beds and occasional basalt lava. The Breiðavík Group was formed over 800 000 years ago (2.0–1.2 million years ago) and, in all, six tillite beds have been identified within the sequence, each representing a glacial stage. Most of the fossil fauna point towards ocean temperatures similar to that of the present, suggesting that the marine deposits were formed during interglacial stages. However, shells of *Portlandia arctica*, a mollusc that lived in then ice-cold waters of the Arctic, have been found on top of two tillite beds. They probably lived at the edge of the ice sheet at Breiðavík during the glacial stages. Six more tillite beds have been identified in the succession above the Breiðavík Group, bringing the total of glacial and interglacial stages recorded at Tjörnes to 14.

Locality 7.3 The Húsavík transform fault

The Húsavík fault is one of two principal faults of the Tjörnes Fracture Zone. The other one, the Kópasker fault, is further to the north. As the name implies, it cuts through the town of Húsavík (see Fig. 7.1). Off shore, this fault forms a valley 5–10 km wide and 3–4 km deep that trends towards the Eyjafjarðaráll graben. The Húsavík fault is characterized by right-lateral fault movement (i.e. standing on one side of the fault, the block on the other side has been displaced to the right). It has been active for about seven million years, during which time the cumulative displacement along the fault has been 60 km.

The fault planes of these fractures are seen on land at the Tjörnes Peninsula as linear northwest-trending structures dissecting the bedrock. One of these fractures passes right through the town of *Húsavík* [66.0446, −17.3394] and is the very source of its existence, because here the displacement of rock masses along the fault has formed a perfect natural harbour. This fault is best viewed from the top of *Mt Húsavíkurfjall* [66.0470, −17.3030], which rises above the town. The fault is easily seen from there as it comes ashore at the harbour and passes through the centre of Húsavík and onwards towards the *Þeistareykir* [65.8869, −16.9758] rift segment southeast of Húsavík, where it connects with the extensional faults of the Northern Volcanic Zone.

Krafla–Mývatn
A paradise of volcanology

The Mývatn area is not only an Eldorado for volcanology enthusiasts and bird watchers, but also has much appeal for those interested in landscape evolution, glacial geology and settlement geography. It is truly an extra-ordinary natural laboratory, because it features almost all aspects of the natural sciences within an area of no more than 15 km^2.

The area takes its name from *Lake Mývatn* [65.6005, −16.9944] (Fig. 7.6), the third largest lake in Iceland (38 km^2), and is situated on the edge of the Krafla volcanic system. The lake is 227 m above sea level and its maximum depth is only 4 m. It sits in a shallow depression, which to the west is bordered by smooth ridges composed of Upper Pleistocene basalts and to the east and southeast by móberg ridges and three large table

Figure 7.6 Aerial view of Lake Mývatn looking north. In the foreground is Skútustaðir; 'rc' indicates locations of rootless cone.

mountains, *Sellandafjall* [65.4088, −17.0392], *Bláfjall* [65.4431, −16.8551] and *Búrfell* [65.5542, −16.6466]. To the north is the basalt caldera volcano, *Krafla* [65.7148, −16.7290], which has produced many eruptions throughout the Holocene. The most recent are the so-called Mývatn Fires (1724–9) and Krafla Fires (1975–84).

The Krafla volcanic system, its fissures and volcanic architecture

The Krafla volcanic system consists of a basalt caldera volcano and an elongate fissure swarm (see Fig. 1.5 and Table 1.4). The Krafla fissure swarm is 10 km wide and 100 km long; it extends from the table mountain, *Sellandafjall* [65.4088, −17.0392], in the south and into the sea at Öxarfjörður [66.2170, −16.7368] in the north. The Krafla central volcano is more than 300 000 years old and consists of a 10 km-wide caldera flanked by a broad and low-lying shield, chiefly composed of basalt-lava flows (see Fig. 7.1). Thus, the volcano is built up from both subglacial and subaerial eruptions. The caldera itself was formed in a large explosive eruption that produced the mixed basalt–rhyolite volcanics of *Mt Hágöng* [65.7400, −16.6861] during the most recent interglacial (about 110 000 years ago). Along outer margins of the caldera are two rhyolite lava domes, *Hlíðarfjall* [65.6774, −16.8566] and Jörundur, which were formed in subglacial eruptions some 15 000–30 000 years ago; they are good representatives of rhyolite table mountains. Within the caldera is a major high-temperature geothermal area, which today is harvested for electricity. Another high-temperature geothermal field, utilized by the diatom-mud processing plant, is situated out on the fissure swarm at the Námafjall Ridge.

The postglacial history of the Mývatn area

During the Younger Dryas stage, the northern margins of the ice sheet were situated along the east–west trending end moraines just north of the *Reykjahlíð* [65.6449, −16.9058] farm (Fig. 7.7). In front (i.e. north) of the Reykjahlíð moraines, glacial rivers formed a small sandur plain that extends north to Mt Hlíðarfjall. Later, this sandur plain has been split up by dislocations (faulting), forming clear exposure through the gravelly glaciofluvial deposits.

After the Younger Dryas glacier receded from the area, subaerial volcanism set in on a larger scale. The Holocene volcanism began with the hydromagmatic eruption that formed the tuff cone *Lúdent* [65.5814, −16.8131] (Fig. 7.7). This eruption episode continued with lava

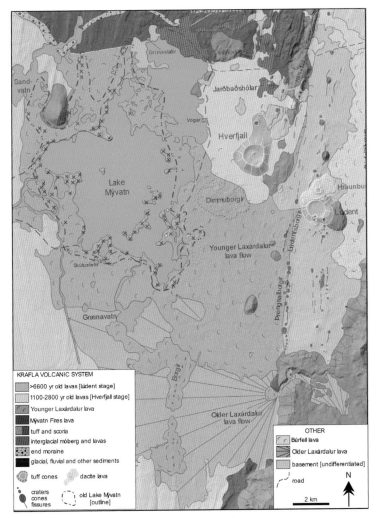

Figure 7.7 Geology of the Mývatn area.

outpourings from volcanic fissures in the Lúdent area and also from short fissures west of the Námafjall Ridge. Lava flows from these fissures are visible just east of the Reykjahlíð farm and on the islet of *Slúttnes* [65.6418, −16.9553]. All of the lava flows from this period pre-date the rhyolite fall deposit (H5) produced by Hekla about 7000 years ago. Near the end of this episode the crater Hraunbunga produced the dacite lava that flanks the Lúdent tuff cone, after which volcanism in the area went quiet for more than 3000 years.

About 3800 years ago, the volcanoes began to rumble again. The first to make its mark was the lava shield *Ketildyngja* [65.4293, −16.6544], about 25 km southeast of Mývatn, which was the first effusive eruption to have a major impact on the landscape evolution in the Mývatn area. This eruption is the source of the Older Laxárdalur lava flow, which entered the Mývatn area through the narrow Selhjallagil gorge (Fig. 7.7). As the lava spread across the basin, it dammed up the outflow from the Mývatn depression and formed the first Lake Mývatn. The lava field known as *Grænavatnsbruni* [65.5059, −16.9633] formed in a fissure eruption about 2200 years ago and is of similar age but slightly older than the Yonger Laxárdalur lava flow field (see below), sits here on top of the Older Laxárdalur flow.

About 3000 years ago the Mývatn area was covered by a rhyolitic tephra fall from the Hekla volcano in South Iceland. This is the so-called H3 tephra layer, which appears as a 5–6 cm-thick white band in the soil profiles around Mývatn. A little later (2800 years ago) a new eruption episode began in the Mývatn area, the Hverfjall–Jarðbaðshólar event, with the birth of the near-perfectly shaped tuff cone, Hverfjall, and the construction of the Jarðbaðshólar scoria complex and associated lava flows. This eruption was followed by a series of fissure eruptions, including the spectacular *Þrengslaborgir* [65.5395, −16.8366]–*Lúdentsborgir* [65.5775, −16.8346] cone row, which formed the Younger Laxárdalur lava flow field about 2200 years ago (Fig. 7.7). Like its predecessor, this lava was a major contributor to sculpturing the landscape in the Mývatn area and is mainly responsible for the scenery as we see it today. It shaped the lake to its present size and formed the maze-like landscape of *Dimmuborgir* [65.5914, −16.9095]. The rootless cones that border the lake shores and form its many islets were formed in this eruption when the lava circumvented the ancestral Lake Mývatn (Fig. 7.6). The rootless cones are truly the trademarks of the scenery at Mývatn and their often symmetrical outlines make them real miracles of nature. Several small fissure eruptions followed the Þrengslaborgir–Lúdentsborgir eruption in the area north of *Hverfjall* [65.6051, −16.8772] and of those the most conspicuous one is the Svörtuborgir ('black fortress') cone row.

Yet again, the volcanoes in the Mývatn area fell silent, this time for just over a thousand years. On 17 May 1724 the silence was broken and the volcanoes began to rumble again. A new period of volcanic activity had begun within the Krafla volcanic system, first with the volcanotectonic episode that became known as the Mývatn Fires and again about 250 years later, with the infamous 1975–84 Krafla Fires.

Locality 7.4 Hverfjall tuff ring

Hverfjall [65.6051, −16.8772] ('crater mountain') is an almost perfectly circular tuff cone formed in a hydromagmatic phase of the Hverfjall–Jarðbaðshólar eruption through the ancestral Lake Mývatn about 2700 years ago. As such, the cone represents the main phase of this event, which has been named the Hverfjall Fires, and is the largest structure on a 5 km-long

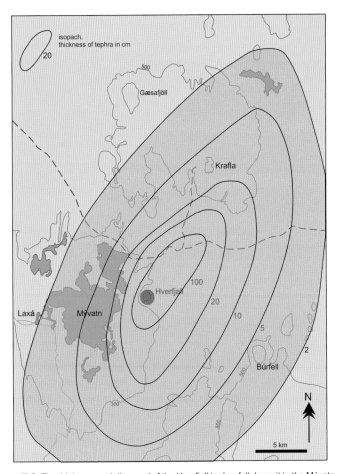

Figure 7.8 The thickness and dispersal of the Hverfjall tephra fall deposit in the Mývatn area.

Figure 7.9 A tree mould in the Hverfjall tephra layer, showing how the tree trunk collapsed from the weight of the tephra plastered onto it by wet pyroclastic surges. The handle of the hammer points in the direction of flow.

volcanic fissure, 90–150 m high and about 1000 m in diameter (Fig. 7.7). It consists of many tephra beds dipping 15–35° in all directions from the crater. These beds are formed in two ways: as fallout from the eruption column and as pyroclastic surges transported by laterally moving collar-like clouds. Away from the cone, the tephra fall was heaviest to the south and northeast of the cone (Fig. 7.8), but the pyroclastic surges were probably dispersed fairly evenly around the cone. North of Hverfjall are excellent outcrops into the tephra deposit, and here the sequence consists of alternating pyroclastic-surge and tephra-fall deposits. Some pyroclastic surges were very wet, because their deposits are plastered, layer upon layer, onto tree trunks. Some trunks were bent to the ground by the extra weight of the tephra (Fig. 7.9). Other surges were hotter and drier, and flowed faster, forming dunes and beautiful cross-bedded deposits in the process (Fig. 7.10). These surges travelled more than 3 km away from the vent.

Jarðbaðshólar [65.6318, −16.8600] is a group of the northernmost and largest (40 m-high) scoria cones on the Hverfjall Fires fissure (Fig. 7.7). The Jarðbaðshólar fissure produced the pāhoehoe and rubbly pāhoehoe lava flow field that covers the ground from the fissures to the eastern shores of Lake Mývatn. In the later stages of the eruption, the lava flowed away

Figure 7.10 Cross-bedded deposits (arrows) within the Hverfjall tephra layer formed by dry pyroclastic surges (the ruler is 30 cm long).

from the northernmost cone in a channel that extends almost halfway to Lake Mývatn. The lava that emerged from these channels formed the rubbly lava that covers the area east of *Vogar* [65.6224, −16.9203]. The Jarðbaðshólar lava flow field covers about 8 km² and its volume is 0.05 km³. The eruption at Jarðbaðshólar took place shortly after the initial phase at the Hverfjall vents ceased, and represents phase two in the Hverfjall Fires.

Locality 7.5
The Þrengslaborgir eruption and origin of Lake Mývatn

The Younger Laxárdalur lava flow was formed by an effusive eruption on the Lúdentsborgir–Þrengslaborgir cone row about 2200 years ago. The cone row runs from the Lúdent tuff cone some 12 km to the south and then shifts to the west onto the Borgir cone row (Fig. 7.7). The fissure's total length is about 16.5 km. Most of the erupted material came up on the central part of the Þrengslaborgir fissure and fed the lava that formed the Dimmuborgir complex and the rootless cones around Lake Mývatn. The lava continued its advance down the Laxárdalur Valley and all the way to the north coast. The total volume of the Younger Laxárdalur is 5–6 km³.

Dimmuborgir is a roughly circular shield-like structure about 2 km in diameter. Forming a natural labyrinth of pillars, some with relict lava tubes, it originated as a secondary lava pond within the Þrengslaborgir lava field, formed when the advance of the flow was hampered by the many rootless

Figure 7.11 The labyrinth of Dimmuborgir showing the pillar-like structures, some with relict lava tubes, left standing after the temporary lava pond drained.

eruptions that occurred when the lava entered the ancestral Lake Mývatn. This is evident in the walls inside the labyrinth, which were constructed by many thin overbank flows spilling out of the pond. When the lava had established dry pathways across the lake, the pond was drained rapidly, leaving behind caved-in and pillar-like structures with relict lava tubes and the filled pathways of ancestral lava rivers (Fig. 7.11).

Locality 7.6 The 1724–9 Mývatn Fires and 1975–84 Krafla Fires

Two major volcanotectonic episodes have occurred in the Krafla volcanic system in the past 250 years: the 1724–9 Mývatn Fires and the 1975–84 Krafla Fires. This activity, especially that of the Krafla Fires, has provided geologists with valuable insights into the mechanics of volcanic systems in Iceland and is therefore summarized here in some detail.

The Mývatn Fires

The volcanotectonic episode referred to as the Mývatn Fires lasted for about five years, from 1724 to 1729. It was characterized by major rifting and faulting on the Krafla volcanic system and was accompanied by a series of explosive and effusive eruptions (Fig. 7.12). The episode began on 17

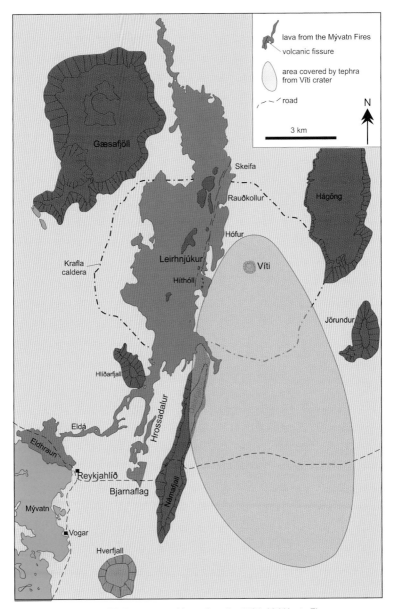

Figure 7.12 The vents and lavas from the 1724–29 Mývatn Fires.

May 1724 with earthquakes, and a hydromagmatic eruption that formed the Víti crater on the southwest slopes of Mt Krafla. Intense earthquakes and large-scale faulting in the area between the Krafla volcano and Lake

189

Mývatn followed this eruption. In January 1726, strong earthquakes were felt again and small explosive eruptions occurred at the Leirhnjúkur hill west of Mt Krafla. Some activity was noted at Bjarnarflag in April that year. On 21 August 1727 a fissure (Skeifa) opened at the northern end of the Krafla caldera, sending lava northeast towards Gæsafjöll. Eight months later, on 28 April 1728, rifting resumed, with lava issued from the Rauð-kollur fissure segment in the Krafla caldera and from short fissures in Hrossadalur, a short distance east of the Reykjahlíð parsonage (Fig. 7.10).

Disturbance resumed in December. On 18 December a large amount of lava belched from the Hófur fissure and from a short fissure at Bjarnarflag, just east of the Reykjahlíð parsonage. Jón Sæmundsson, the vicar at Reykjahlíð described this volcanic outburst in Bjarnarflag as follows:

> In some places the fire had greater action with huge and great flames, and continuously threw up all around itself glowing rocks and stones, which are so wet [fluid] that they, still glowing, flowed like melted copper. This ejection of glowing rocks forms large crater-like hills or rocky hills [where the fire emerges], which are hollow inside with circular rims. Inside these source pits of the fires were an intensive and strong *inflammatio ignis* with such a strong storm that the flames were thrown high up into the sky. The same fire threw up continuously glowing and wet rocks that when they hit the ground, they flowed like running water and burned up everything in their way.

It is obvious that Jón Sæmundsson is describing lava fountains, and his depiction of glowing and 'wet' rocks thrown up into the air and flowing like water after hitting the ground is the oldest known description of a fountain-fed lava flow. However, Krafla was not done and another eruption started on 30 June 1729. This time a fissure opened at Hítarhóll just south of Leirhnjúkur, sending a large volume of lava down towards Reykjahlíð and all the way into Lake Mývatn (Fig. 7.12). This flow ruined four farms, including the Reykjahlíð farm, and forced the vicar and his family to seek shelter at Skútustaðir. However, the lava did not manage to destroy the Reykjahlíð church, which stood on a small hillock and still does, and now is completely surrounded by lava. Seventeen years later, in July 1746, a small explosive eruption occurred at Leirhnjúkur, suggesting that this volcanotectonic episode at Krafla may have lasted for up to 22 years. The cumulative length of the volcanic fissure formed in the Mývatn Fires is about 13 km.

The Krafla Fires

The volcanotectonic episode of the Krafla Fires began in December 1975 and lasted until September 1984. The first sign of renewed activity within the Krafla volcanic system was an increase in earthquake activity in June 1975, gradually building up to earthquakes of magnitude 4 on the Richter scale. The first eruption broke out from a short fissure at Leirhnjúkur, producing a small lava flow and a few explosions (Fig. 7.13). Initially, the earthquake activity was confined to the region below the caldera, but it spread out quickly along the fissure swarm to the north coast at Öxarfjörður and south towards Lake Mývatn. This rifting was associated with a 1.5 m widening of the fissure swarm, and a 2 m-deep graben was formed along the central part of the fissure system. The eruption was followed by an immediate subsidence within the caldera floor, forming a bowl-shaped depression. This was followed by a gradual inflation of the caldera floor over the next few months. This pattern of gradual inflation and sudden deflation recurred several times in the following years. Each cycle began with inflation of the caldera floor at a rate of 7–10 mm/day. When the inflation reached a critical level, the earthquake activity increased steadily until another sudden deflation event occurred. This is known as a rifting event. During the Krafla Fires, the rifting began by failure of the crust within the caldera and migrated from there along the fissure swarm, as indicated by the outward-propagating earthquake swarms. Magma flowed into and filled the open cracks (see Fig. 1.6b) and in some cases found a pathway to the surface to produce an eruption.

In total, 21 rifting events occurred during the 1975–84 Krafla volcanotectonic episode. The first four rifting events occurred in October 1976, January, April and September 1977, but only the last two deflation events resulted in an eruption at the surface. Other notable rifting events occurred, for example, in March, July and October 1980, November 1981 and September 1984. Many of the later deflation events resulted in fairly large lava eruptions (Fig. 7.13). After the 1984 event, inflation resumed within the caldera until it ceased in early 1985. Minor, intermittent inflation/deflation cycles were recorded between 1986 and 1989; since then the caldera floor has been subsiding at a slow but steady rate.

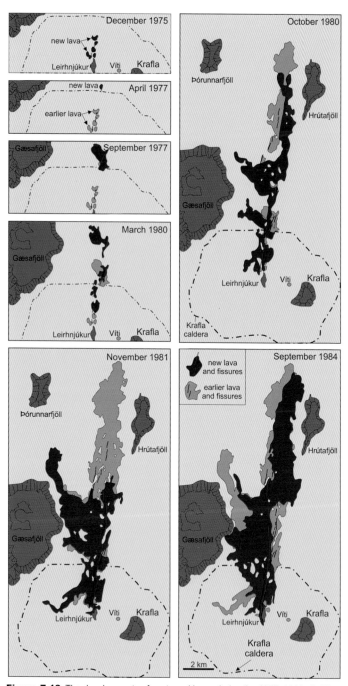

Figure 7.13 The developments of vents and lavas during the 1975–84 Krafla Fires.

The total widening of the Krafla fissure swarm during the nine-year volcanotectonic episode was in the order of 900 cm and the widening during the 1724–29 Mývatn Fires was of the same magnitude. As you might recall, the long-term average spreading rate of the plate boundary in Iceland is about 2 cm/year. Thus, through rifting in two relatively short volcanotectonic episodes spanning 15 years in total, the volcanic system has made up for the previous 1000 years, which were apparently marked by little or no spreading on this particular segment of the plate boundary. This evidence seems to suggest that although the spreading along the plate boundary is a continuous process, the rifting of the brittle crust is periodic and at any time confined to episodes, such as the Mývatn and Krafla Fires, on individual volcanic systems.

19 October 1980 eruption at Krafla volcano. Photo by Páll Imsland.

Chapter 8

The north and northwest

General overview

North and Northwest Iceland are composed entirely of rocks belonging to the Neogene Basalt Formation, except for the outlier and extinct Skagi volcanic belt which was formed during the Plio-Pleistocene time (Fig. 8.1). Similar to East Iceland, these regions are characterized by rugged mountains and deep fjords, a landscape of extensive glacial erosion.

The age of the Neogene Basalt Formation in North Iceland ranges from about 12 million years in the deepest part of the stratigraphical column at the northern tip of Tröllaskagi to about 6–7 million years in the region from Blönduós across the Central Highlands to the bottom of Eyjafjarðardalur. At the *Skagi* [65.8936, −20.0580] peninsula the much younger (<3 million years) Skagi volcanic belt is superimposed on the eroded Neogene Basalt Formation (Fig. 8.1). Even older rocks are exposed at the Vestfirðir Peninsula, where some, 18 million years old, occur deep in the stratigraphical column in mountain regions immediately south of Ísafjarðardjúp. The succession becomes progressively younger towards the southeast and at the neck of the peninsula; the age is about eight million years, younging to six million years on the south side of Hvamsfjörður (Fig. 8.1).

The construction of the Neogene Basalt Formation in North and Northwest Iceland was identical to that of East Iceland. Altogether, 23 extinct volcanic systems have been identified in the western part of the Neogene succession, which also contains seven major sedimentary horizons (labelled 1–7 on Fig. 8.1), which are, from west to east, the *Selárdalur* [66.1226, −23.4394]–Botn (1; 15 million years old), *Saurbær* [65.4773, −24.0024] (2; 13 million), *Brjánslækur* [65.5261, −23.1924] (3; 12–11 million), *Bjarkarlundur* [65.5573, −22.1145] (4; 11–10 million), *Húsavíkurkleif* [65.6430, −21.6342] –Tröllatunga (5; 10 million), Tindar–*Mókollsdalur* [65.5214,

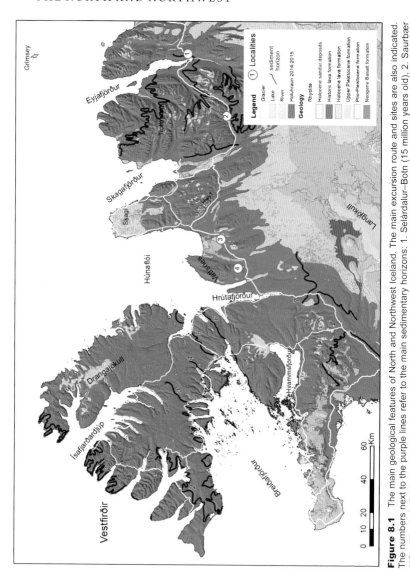

Figure 8.1 The main geological features of North and Northwest Iceland. The main excursion route and sites are also indicated. The numbers next to the purple lines refer to the main sedimentary horizons: 1. Selárdalur–Botn (15 million years old), 2. Saurbær (13 million), 3. Brjánslækur (12–11 million), 4. Bjarkarlundur (11–10 million), 5. Húsavíkurkleif–Tröllatunga (10–9 million), and 6. Tindar–Mókollsdalur (8 million).

−21.5301] (6; 9–8 million) and an unnamed one at the bottom of Hrútafjörður (7; 7–6 million). The sediments represent fluvial, lake or peat-bog (lignite) deposits and are all formed on land.

These sedimentary horizons contain fossils or fossil imprints of the plants that lived at the time. For example, the 15 million-year-old Selárdalur–Botn horizon, which extends from Selárdalur to *Kögur* [65.9140, −23.8316]

(Fig. 8.1), contains 20 m-thick lake sediments with leaf imprints and pollen (seeds) of trees such as vine, walnut, elm, beech, dawn redwood and pine. This fossil fauna most resembles the eastern deciduous forest of North America, showing that at this time the climate of Iceland was similar to that of the southeastern USA today. In general, fossil fauna indicate a gradual and steady cooling of the climate from 18 to 6 million years ago, which is consistent with similar evidence elsewhere. Each of these sedimentary horizons can be followed for tens of kilometres, and they appear to represent a break in the build-up of the volcanic succession in this region, which can either be due to a shift in location of the volcanic activity (as it moves further away) or to reduction in magma output. This allowed for thick sedimentary sequences to accumulate without interruption. The apparent 1–2 million-year cyclicity in their occurrence is interesting but has yet to be explained adequately. A significant structural change to the Neogene Basalt Formation has been documented across one of these sedimentary horizon, namely the 15 million year old Selárdalur–Botn. Above the horizon the formation is characterized by a shallow regional dip (5–10°) to the southeast, whereas below it the dip is about 5° to the west (see Fig. 1.2). This reversal in the regional dip implies different source regions for the successions above and below the Selárdalur horizon. The southeast dip of the overlying succession is consistent with an origin within the now-extinct Snæfellsnes–Vatnsdalur Volcanic Zone, whereas the westward dip of the succession below the horizon implies origin within a volcanic zone that is now buried in the sea somewhere off the northwest coast of Iceland. By the same token, this implies that eastward shifts of the volcanic rifts have been taking place throughout Iceland's short geological history.

There is one excursion in this region, which follows Highway 1 across North Iceland from Fnjóskadalur to Hrútarfjörður (Fig. 8.1).

Fnjóskadalur to Skagi

Locality 8.1
Fnjóskadalur valley and the ice-dammed lake sediments

The *Bleiksmýrardalur* [65.4050, −17.7748] Valley and its northward extension, the *Fnjóskadalur* [65.7516, −17.8839] Valley, form the longest valley system in Iceland, 150 km in total. Situated within the Neogene Basalt Formation, it was originally formed by fluvial erosion during the late

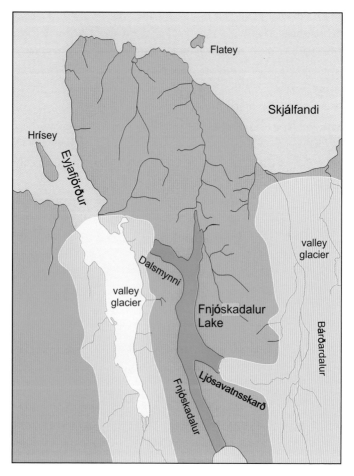

Figure 8.2 The condition during Allerød Stage when the ice-dammed lake filled the Fnjóskadalur valley. Also shown are the positions of the main valley glaciers during the Allerød stage.

Neogene period and was later carved out by the Ice Age glaciers. Scattered outcrops into young Upper Pleistocene lavas show that lava flows came down this valley system during one of the interglacial stages and at least one advanced all the way into Eyjafjörður (Fig. 8.1). During the waning stages of the Weichselian glaciation (i.e. Allerød stage), the Fnjóskadalur valley was occupied by an ice-dammed lake, which was kept up by the valley glacier that filled Eyjafjörður (Fig. 8.2). The gravel terraces that circumscribe the lower valley slopes were formed in this lake. The so-called Skógar tephra, a pale pumice-fall deposit of rhyolite composition, is found in these terraces. The formation of this tephra has been linked to a large explosive Plinian eruption in the Mýrdalsjökull volcano at about 12 000 years ago (see p. 127).

On the way

The extinct Öxnadalur [65.5873, −18.5300] central volcano, which was active 8–9 million years ago, is located southwest of the town of Akureyri, where Highway 1 climbs out of the Öxnadalur Valley. In its time the volcano rose to about 1000 m above the surroundings and was capped by a 6 km-wide and 1 km-deep caldera. An exceptional exposure of the eastern caldera wall is clearly visible in the top cliff face of *Mt Hólafjall* [65.5344, −20.7064] above the *Engimýri* [65.5715, −18.5350] farm (Fig. 8.3). Following the distinctively layered basalt-lava series that forms the outer slope of the volcano from the eastern end of Mt Hólafjall, it is suddenly terminated by a near-vertical caldera wall in the western side of the mountain. Here, irregularly bedded sediments and volcanics fill in the caldera. The volcanics consist of cube-jointed basalt flows, pillow lavas, and móberg tuffs and breccias, indicating that at some stage the caldera contained a lake.

The pile of debris that forms the hills at the foot of the mountains on the opposite side of the valley is called Hraun and was formed as a rock glacier during early postglacial times. The rock glacier blocked the small valley above and a small lake formed in the depression behind its terminus. A little farther west, near the deserted Bakkasel farm at the bottom of

Figure 8.3 The caldera margins (arrowed) of the Öxnadalur volcano, with the lava flow-dominated east flank of the volcano to left of arrow and the more irregularly stratified caldera fill to the right.

Öxnadalur, a large rock called Lurkasteinn stands proud to the south of the road. It is mentioned in one of the Icelandic sagas (i.e. Þórðar sögu hreðu) as the scene of a duel between Þórður the squabbler and Sörli the sturdy one, which resulted in Sörli's death.

Locality 8.2   Kotagil

The road across Öxnadalsheiði pass descends into *Norðurárdalur* [65.4457, −19.0070], a narrow glacially carved valley with an almost perfect U-shaped cross section. On the north side of Norðurárdalur, the small tributary stream *Kotá* [65.4424, −19.0366] flows through an impressive gorge, which has been cut into the Neogene Basalt Formation by recurrent but infrequent flash floods.

In the west wall of the gorge a little way up from its base, about 200–300 m from its mouth, there are moulds of ancient tree trunks at the contact between two basalt-lava flows (Fig. 8.4). The presence of such tree moulds has long been noted in Neogene lava flows in Iceland and can be distinguished from other voids by a boxwork pattern on their inner surface, or by the branching form of the mould itself. In pāhoehoe lavas, tree moulds are typically found near the base or top of a flow lobe and, as in *Kotagil*

Figure 8.4 Tree moulds at the point of contact between two pahoehoe lobes at Kotagil in North Iceland.

[65.4544,−19.0569], there is never any soil or sediment between the two lobes. Therefore, it is unlikely that the trees were standing on the lower lava when they were alive and growing. So, how did the tree trunks assume a position where they were buried and burned by the lava to form the moulds? We know from observations in Hawaii that coconut palms are burned and toppled onto the surface of inflating pāhoehoe flows. The trees are preserved where subsequent lava lobes cover the log before it decays. The Kotagil and tree moulds in Icelandic lavas were most probably formed in a similar way.

On the way
As you follow Highway 1 along the southern slopes of *Bólstaðarhlíðarfjall* [65.5338, −19.8323], just before entering the *Langidalur* [65.5735, −19.9996] valley en route to Blönduós, a keen observer might notice the many dislocations visible within the westward-dipping strata. These are classic examples of normal faults, where one side has been displaced downwards relative to the other.

Locality 8.3   Vatnsdalshólar landslide formation

Vatnsdalshólar [65.4956, −20.3778] is a chaotic cluster of hummocks and hills that stretch across the mouth of the Vatnsdalur Valley (Fig. 8.5). These mounds and hummocks are made entirely of assorted rock debris and are one of the best examples in Iceland of deposits produced by a rock avalanche. Here, the flow-like shape and the hummocky surface morphology that characterize deposits of this type of are perfectly preserved (Fig. 8.5a). The avalanche that formed Vatnsdalshólar occurred more than 7000 years ago. The scarp left behind by the collapsing rock mass can still be seen in the western slope of *Mt Vatnsdalsfjall* [65.4592, −20.2492] at about 900 m above sea level (Fig. 8.5b). As the avalanche hurtled down the slope, it picked up momentum and ran up the opposing slope to a height of about 70 m. It also spread debris up to 4 km away from its source and covered an area of about 15 km².

Two smaller rock avalanches have descended from Mt Vatnsdalsfjall in historical times. The first one cascaded down the mountain slopes in 1545 and ran over the Skíðastaðir farm, killing 14 people. The small lake, Hnausatjörn, just south of the highway, is formed in the debris from this avalanche. The second one occurred in 1720 and surged down the slopes

Figure 8.5 (a) The Vatnsdalshólar debris avalanche deposits by the rest stop on Highway 1 in front of the Vatnsdalur valley. (b) The debris avalanche viewed up-flow (i.e. to the east). The arrow indicates the flow direction, the avalanche scar is right above the arrow, while the debris avalanche deposit is in the foreground.

just south of Hnausatjörn. The debris dammed the river and formed the larger lake, Flóðið. Six people were killed in this event.

Locality 8.4 Vatnsnes syncline and the Skagi volcanic succession

In approaching the *Vatnsnes* [65.5866, −20.7963] Peninsula, note that the regional dip of the strata has reversed. At Mt Vatnsdalsfjall and other mountains to the east, the strata dip at 5–10° west or west-northwest, whereas at Vatnsnes and the mountains farther to the west the strata generally dip to the east. The strata with these opposing dips converge on the east side of Vatnsnes, defining a north-trending axis of a syncline (see Fig. 1.2). This is the Vatnsnes syncline, which represents an old rift zone that became inactive about five million years ago and is the precursor to the currently active

North Volcanic Zone. Prior to five million years ago, the northern limb of the actively spreading plate boundary across Iceland was situated at the Vatnsnes rift zone. The three extinct central volcanoes that stretch from Mt Vatnsdalsfjall across to Vatnsnes were formed by volcanism on this rift zone. The much younger (< 3 million years old) Skagi volcanic succession (Fig. 8.1) was formed in a rejuvenation stage on this old rift zone – similar to recent volcanics on the Snæfellsnes Volcanic Belt.

When crossing the *Holtavörðuheiði* [64.9655, −21.0621] mountain road on a bright day, a large, steep-sided mountain capped by a cupola-shape glacier appears on the horizon to the southwest. This is *Eiríksjökull* [64.7734, −20.3963] (1675 m), the largest table mountain in Iceland. The basal diameter of the mountain is about 10 km and it rises more than 1000 m above its surroundings. The lowest 350–500 m consist of móberg tuffs and breccias, which are capped by a > 700 m-tall lava shield. It was presumably formed in a single subglacial eruption some time during the most recent glacial stage.

The mountains Kirkjufell and Brimálarhöfði near Grundarfjörður, Snæfellsnes peninsula. Photo by Ármann Höskuldsson.

Chapter 9

The west

The Snæfellsnes Peninsula is about 80 km long but only 10–30 km wide (Fig. 9.1). A 700–1000 m-high mountain chain forms the backbone of the peninsula, which at its tip is crowned by the Snæfellsjökull stratovolcano.

Snæfellsnes–Borgarfjörður
Volcanic zones and rift jumps
Snæfellsnes Volcanic Belt – the mantle-plume trail
The mountain chain that forms the crest of the Snæfellsnes Volcanic Belt is the product of outlier volcanism that began about a million years ago along the trail of the mantle plume. These young volcanics rest unconformably on the 1.2 million-year-old *Búlandshöfði* [64.9424, −23.4758] sedimentary horizon, which in turn rests on the much older and eroded bedrock of the Neogene Basalt Formation (Fig. 9.1). The Snæfellsnes Volcanic Belt comprises three volcanic systems trending west-northwest, which exhibit a subparallel (en echelon) arrangement along the peninsula. These are, from east to west, the *Ljósufjöll* [64.9145, −22.5803], *Lýsuskarð* [64.8520, −23.2075] and *Snæfellsjökull* [64.8060, −23.7768] volcanic systems (see Fig. 1.5). The small cluster of Upper Pleistocene volcanics at Snjófjöll may represent the fourth system of this belt.

The 90 km-long and 20 km-wide Ljósufjöll system is the largest of the three systems and has been active for at least 0.7 million years. Its centre is at the multi-coloured Ljósufjöll mountain complex, which consists of andesite to rhyolite volcanics, mostly formed by subglacial eruption during the Upper Pleistocene. During the Holocene, the activity has been characterized by the basaltic volcanism that extends across the system, from the fissures of *Berserkjahraun* [64.9640, −22.9640] in the west to the *Grábrók* [64.7700, −21.5375] scoria cone in the east. In all, 23 basalt eruptions have occurred in the system in the past 11,500 years, including one historical

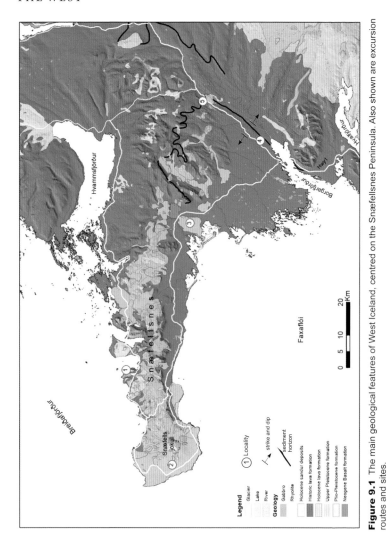

Figure 9.1 The main geological features of West Iceland, centred on the Snæfellsnes Peninsula. Also shown are excursion routes and sites.

eruption and the well-known spatter-cone volcano, *Eldborg á Mýrum* [64.7956, −22.3208] (see locality 9.3). The latter is the source of the term 'eldborg', commonly used in Iceland for spatter cones.

The Lýsuskarð system is located in the Central Highlands of the peninsula and is the smallest one, 30 km long and 5 km wide. The core of this system is a central volcano that began to erupt about two million years ago, but was most productive after one million years. Only two eruptions are known within the system in the past 10 000 years, both basalt-lava eruptions, one forming magnificent lava falls that cover the southern slopes of the highlands near Lýsuhóll.

The third volcanic system, Snæfellsjökull, which is named after the infamous stratovolcano that sits on the western tip of the peninsula, is about 30 km long and has been active for just over 700 000 years. Although the Snæfellsjökull system has been very active during the Holocene, with a tally of more than 25 effusive basaltic and three explosive Plinian eruptions, none have occurred in the past 1100 years, the time of human settlement in Iceland (see also locality 9.2).

The Neogene geology: extinct rift zone and rift jump
The Neogene geology of western Iceland tells an intriguing story of a rift jump, because it dictates how plate boundaries were shifted from one volcanic zone to another. Unfortunately, the geology is complex and thus cannot be unravelled without involving demanding concepts, so please bear with us.

The youngest rocks of the Neogene Basalt Formation at Snæfellsnes are about six million years old. They occur along the axis of the Snæfellsnes syncline, which stretches obliquely across the peninsula from *Garðar* [64.8037, −22.2608] to Álftafjörður [65.0169, −22.6461], where it turns eastwards across *Hvamsfjörður* [65.0747, −22.0631] (Fig. 9.1). On either side the strata dip towards the syncline axis, and the rocks become progressively older away from it. Northwards, the Neogene succession can be followed without a major break all the way across the Vestfirðir Peninsula or back to rocks that were formed about 18 million years ago (see Chapter 8). To the south they can be followed across the Mýrar region before the Neogene rocks disappear beneath the much younger products of the West Volcanic Zone on the south side of Borgarfjörður (Fig. 9.1).

The Snæfellsnes syncline is an ancient rift zone and was characterized by volcanism identical to that found in the currently active volcanic zones; it hosted at least four active volcanic systems in its prime. This rift zone was an active spreading centre until six million years ago, when the plate boundary shifted to its current position within the West Volcanic Zone. It is a continuation of the extinct Vatnsnes rift zone and it represents the southern limb of the plate boundary across Iceland as it was prior to six million years ago.

Across the imaginary line extending northeast betwen localities 4 and 5 (Fig. 9.1), the dip of the Neogene strata reverses again. North of this line the strata dip to the northwest, whereas on the south side they dip to the southeast. This is the *Borgarfjörður anticline* [64.6347, −21.7911] (see locality 9.4). The oldest rocks in Borgarfjörður are exposed in the axis of the anticline and are 13 million years old. Away from the anticline axis the rocks become younger: 6–8 million years old at Snæfellsnes and *Hítardalur* [64.8278, −22.0651], and about 12 million years old where the Neogene basalts disappear under the younger products of the West Volcanic Zone to the south.

In the Neogene strata on either side of the Snæfellsnes syncline are thick conspicuous sedimentary units, the Tindar–Mókollsdalur and Hreðavatn horizons (Fig. 9.1), which were formed during a period when there was a lull or shift in the loci of volcanic activity. The Tindar horizon is about eight million years old and occurs within the succession without a major time break. The story is quite different for the Hreðavatn horizon. The Hreðavatn sedimentary horizon formed about 6–7 million years ago and can be followed from *Hítarvatn* [64.8830, −21.9605] east across to *Hreðavatn* [64.7588, −21.5812] and then southwest to *Mt Hafnarfjall* [64.4833, −21.9042]. At Hítarvatn, the locality closest to the Snæfellsnes syncline, the age difference between the lavas on either side of the Hreðavatn horizon is very small (< 0.3 million years). Following the Hreðavatn horizon to the south and southwest, the age difference between the lavas below and above increases steadily. At Hreðavatn the difference is about five million years, the underlying lavas being 12 million years old, whereas the ones on top are 6.5 million years old. At Mt Hafnarfjall the difference is even greater at seven million years. This age sequence tells us two things.

First, the age progression from Hítarvatn to Hreðavatn and Mt Hafnarfjall shows that the Neogene volcanics become progressively older across the Mýrar region. This is to be expected if they were formed by volcanism in the now extinct Snæfellsnes rift zone and subsequently moved from it by spreading to the southeast. By the same token, the equal age of the lavas at Hreðavatn and Mt Hafnarfjall is expected because they are at the same distance from the spreading axis. Thus, moving southwards, the Hreðavatn sediments were blanketing progressively older and more eroded lava sequences. Consequently, away from the Hítardalur locality, the contact between the Hreðavatn sediments and the underlying Neogene lavas is an unconformity, because part of the Neogene succession is missing. Secondly, the volcanic formations immediately above the sedimentary horizon become gradually younger from Hítarvatn across to Hreðavatn and Mt Hafnarfjall,

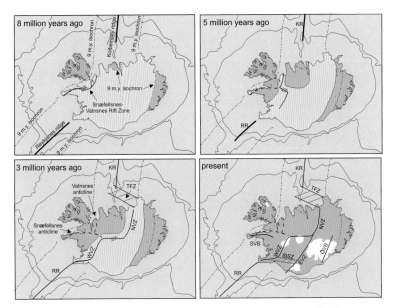

Figure 9.2 Evolution of rifts and volcanic zones in Iceland during the past 16 million years.

indicating that seven million years ago volcanism had begun in the area to the east of Hreðavatn and gradually spread to the southwest over the next two million years. At the same time the activity was gradually declining in the Snæfellsnes rift zone and the last volcano died out about six million years ago. This sequence clearly indicates that the centre of volcanism and spreading was shifting from the Snæfellsnes rift zone to the current West Volcanic Zone (see Fig. 1.3), a rift-zone jump that took at least two million years to complete. The Borgarfjörður anticline was formed as a result of this shift. First, the strata were tilted to the northwest because of loading of the crust by the volcanism at the Snæfellsnes rift zone. Later, the part south of the anticline axis was tilted to the southwest by loading within the West Volcanic Zone.

At present, this same process was in progress between the active West and East Volcanic Zones. The northwesterly drift of the North America/ Eurasia plate boundary relative to the Iceland mantle plume is the driving force behind these rift jumps. The history of these plate-boundary shifts and evolution of the rift zones in Iceland is illustrated in Figure 9.2.

Locality 9.1 Búlandshöfði interglacial sediment sequence

Towards the base of the mountains on the north side of the Snæfellsnes Peninsula is a 20–50 m-thick sedimentary sequence that can be followed for

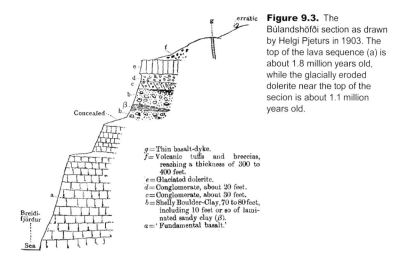

Figure 9.3. The Búlandshöfði section as drawn by Helgi Pjeturs in 1903. The top of the lava sequence (a) is about 1.8 million years old, while the glacially eroded dolerite near the top of the secion is about 1.1 million years old.

g = Thin basalt-dyke.
f = Volcanic tuffs and breccias, reaching a thickness of 300 to 400 feet.
e = Glaciated dolerite.
d = Conglomerate, about 20 feet.
c = Conglomerate, about 30 feet.
b = Shelly Boulder-Clay, 70 to 80 feet, including 10 feet or so of laminated sandy clay (β).
a = 'Fundamental basalt.'

30 km from *Grundarfjörður* [64.9241, −23.2401] to Ólafsvík [64.8947, −23.7059]. Good cross-sectional views through the sediments are found in outcrops at *Mt Stöð* [64.9561, −23.3659] and *Mt Búlandshöfði* [64.9428, −23.4888], from which the sedimentary sequence takes its name (Fig. 9.3). The Búlandshöfði sediments and overlying volcanic formations record two glacial stages and one interglacial stage that occurred 0.9–1.2 million years ago.

At Stöð and Búlandshöfði the lavas of the Neogene Basalt Formation rise to 130 m above sea level. The upper surface of the lava pile is striated, showing that the Neogene Formation has been scraped and eroded by an advancing glacier. The tillite bed mixed with marine sediments rests unconformably on this surface. Mollusc fossils (e.g. *Portlandia arctica*) have been found in these marine sediments, which indicate an ice-cold sea. This sequence represents the first glacial stage recorded by the Búlandshöfði sequence. On top of these cold-sea sediments are clay-rich marine deposits containing fossils of molluscs that live in the ocean around Iceland today (i.e. *Arctica islandica*, *Mytilus edulis* and *Nucella lapillus*). These fossils show that the sea temperature and the climate had become much warmer when these sediments were laid down. The ice sheet of the previous glacial stage had receded and the region now enjoyed the warmer climate of an interglacial stage. These warm marine sediments are missing in the Mt Stöð section. In their place is a unit of fine-grained sandstone representing a delta built into a lake. This sandstone contains many leaf imprints and pollen grains of alder, willow and birch: the land was fully vegetated.

Figure 9.4 Snæfellsjökull volcano. View to west.

Shortly afterwards the Búlandshöfði sediments were covered by a series of lava flows formed by fissure eruption within the rejuvenated Snæfellsnes volcanic zone about 1.1 million years ago. The upper surface of this lava sequence is polished and striated, showing that the ice sheet of the subsequent glacial stage advanced over these lavas. The subglacial móberg ridge

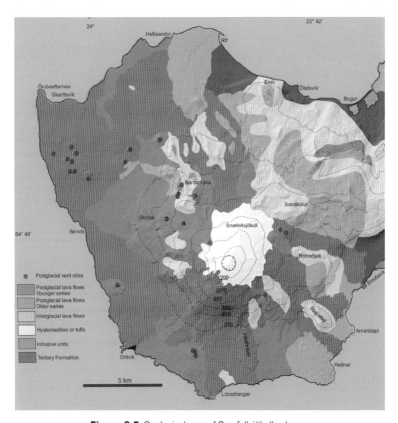

Figure 9.5 Geological map of Snæfellsjökull volcano.

that caps the succession was built in an eruption beneath this ice sheet and it records the second glacial stage in the Búlandshöfði sequence.

Locality 9.2   Snæfellsjökull central volcano – a journey to the centre of the Earth

At the tip of the Snæfellsnes Peninsula is the perfectly cone-shape *Snæfellsjökull* [64.8060, −23.7768] stratovolcano (Fig. 9.4), crowned by a 1.1 km-wide summit caldera. The volcano is 1446 m high and on a clear day it is clearly visible from distant places such as Reykjavík. Because of its majestic stature, it was for long thought to be the highest mountain in Iceland (i.e. before accurate mapping). However, the volcano is perhaps best known for being the starting point of the epic journey portrayed by Jules Verne in his classic adventure story, *A Journey to the Centre of the Earth*.

Snæfellsjökull volcano has been active for the past 840 000 years. In postglacial times it has produced more than 25 eruptions, all of which are prehistoric (Fig. 9.5). Three of those were Plinian eruptions, which occurred at ~9000, 3700 and 1810 years ago and dispersed rhyolitic tephra over parts of the Snæfellsnes Peninsula (Fig. 9.6). The remainder were mafic lava-producing eruptions emanating from vents in the summit region as well as parasitic ones at the foot of the volcano (Fig. 9.5).

Figure 9.6 A vertical profile through the silicic pumice fall deposit formed the 1810 PB Plinian eruption at Snæfellsjökull. Site of section is at the foot of the eastern flank of the volcano.

Sn-I tephra

soil

Locality 9.3   Eldborg á Mýrum – a spatter-cone volcano

> That night an earth fire broke out and the Borgarhraun lava was
> formed. The farm stood where the cone is standing now.

This succinct description is all that was written in the Book of Settlement about the eruption that occurred in the *Mýrar* [64.6782, –22.0753] region in about AD 900. Initially, it was thought that this description referred to the formation of the fort-like spatter cone Eldborg (at Mýrar) and its lava flow. Later studies have shown that it is not so, because the eruption that formed Eldborg occurred more than 5000 years ago. However, the description is not entirely fictional because **tephrochronology** shows that the nearby *Rauðhálsar* [64.7462, –21.8842] cone and lava flow were formed in an eruption shortly after the first settlers arrived in Iceland.

The Eldborg lava is the largest (33 km²) of six small lava flow fields that cover the lowlands of Mýrar and were formed by eruptions on short west-northwest trending fissures within the Ljósufjöll volcanic system. Eldborg is the largest cone of five that delineate a 1 km-long fissure. It is a complete spatter cone and it would be difficult to find a better example (Fig. 9.7). The cone is ellipsoidal in plan view, 250 m long and 180 m wide, and rises 50 m above its surroundings. The cone rims are steep (40–60°) and on the outside

Figure 9.7 Aerial view of the spatter cone Eldborg at Mýrar.

they are draped by small fountain-fed lava streams. During the eruption a small lava pond occupied the crater, and overflows from this pond formed the many 2–6 cm-thick lava units exposed in the inner cone walls. Most of the lava was transported away from the vents in subsurface lava tubes to form the bulk of the flowfield produced by the Eldborg eruption. However, the darker and less vegetated lava apron immediately around the cone is composed of fountain-fed flows formed during the later stages of eruption.

Locality 9.4   The Borgarfjörður anticline

Travelling east on Highway 1 away from the town of *Borgarnes* [64.5530, −21.9003] and towards *Bifröst* [64.7660, −21.5537], take note of the dip direction of the basalt-lava flows poking through the vegetation cover on either side of the road. North of the road the lava beds dip to the north-northwest, whereas on the south side they dip to the southeast. This reversal of dip direction across the road is explained by the fact that here it follows the axis of the Borgarfjörður anticline (see Fig. 9.1).

Locality 9.5   The Grábrók scoria cone volcano and the Hreðavatn unconformity

Grábrók is the largest cone on a 2 km-long scoria cone row. The eruption that formed the Grábrók scoria cone and the surrounding lava flow occurred 3600 years ago and represents the easternmost volcanic fissure in the Ljósufjöll volcanic system. Ready-made trails on the outer slopes of Grábrók make it quite suitable for a hike to the top, which provides an excellent view over the fertile regions of Borgarfjörður.

Nearby in the Brekkuá gorge is an outstanding exposure of the Hreðavatn sedimentary horizon and the associated unconformity. Here the 6–7 million-year-old Hreðavatn sediments consist of fossiliferous siltstone and sandstone beds, which rest unconformably on an eroded and southeast-dipping (10–15°) lava-flow sequence formed some 12–13 million years ago (Fig. 9.8). The lava sequence above the Hreðavatn horizon rests conformably on the sediments, and small pillow-like lobes occur at the base of the lowest lava, implying that it was emplaced into standing water.

On the way

Skorradalur, 2–3 km wide and 25 km long, is a classic example of a U-shaped glacial valley. Lake *Skorradalsvatn* [64.5172, −21.4851] (15 km long, 43 m

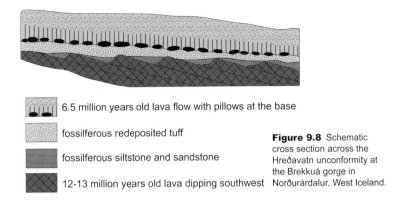

 6.5 million years old lava flow with pillows at the base

fossilferous redeposited tuff

fossilferous siltstone and sandstone

12-13 million years old lava dipping southwest

Figure 9.8 Schematic cross section across the Hreðavatn unconformity at the Brekkuá gorge in Norðurárdalur, West Iceland.

deep) sits in the glacially eroded depression behind a threshold of solid rock at the valley mouth. Initially, the valley was filled to the brim with glacial ice, which was reduced to a valley glacier as the climate became warmer during the waning stages of the Weichselian glaciation. As the glacier retreated farther about 15 000 years ago, the valley was inundated by the ocean, as is indicated by the fossil shells found in marine sediments at the valley mouth. Marine sediments of the same age are also found along the coast at Melabakkar to the west of Mt Hafnarfjall. There the sediments are capped by tillite formed as end moraines in front of the Younger Dryas glacier. This sedimentary sequence indicates that the Borgarfjörður region was below sea level for more than 2000 years during late glacial times.

Mt Hvalfell [64.3857, −21.2125] is a small table mountain at the bottom of Hvalfjörður, formed in a subglacial eruption more than 15 000 years ago. It completely fills in the valley, such that, when the glacier retreated, a small lake was formed in the valley above the mountain (Fig. 9.9). Lake Hvalvatn is about 160 m deep, the second deepest lake in Iceland.

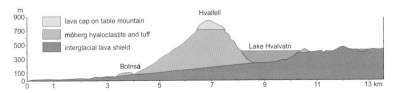

Figure 9.9 Cross section of Botnsdalur in Hvalfjörður, West Iceland. Lake Hvalvatn was formed when the table mountain Hvalfell was built by a subglacial eruption towards the end of the Weichselian glaciation, thus forming a natural dam across the valley.

215

Aerial view of the Askja volcanoe. Photo by Oddur Sigurðsson.

Chapter 10

The Highlands

General considerations

We present the two highland excursions in a succession from north to south or vice versa; routes and places mentioned in the text are shown in Figures 5.1 and 7.1. Each of these excursions can also be undertaken as separate tours. For example, *Askja* [65.0348, −16.7592]–*Kverkfjöll* [64.6874, −16.6028] can be done individually with Lake Mývatn as a starting point and similarly the *Landmannalaugar* [63.9889, −19.1796]–*Veiðivötn* [64.1419, −18.7662] excursion can be undertaken from any of the major centres in South or Southwest Iceland. However, before delving into the descriptions, we would also like to offer a few words of advice. A trip across the Central Highlands in Iceland is not to be treated lightly and one should be prepared to face the elements in this harsh environment. Travel into the highlands without the guidance of experienced personnel is not recommended, other than for persons with serious mountaineering experience. There are no roads, only tracks, and only a few of the rivers have been bridged. Stock up with provisions before you leave, because there are no shops in the region.

The journey to Askja and the Kverkfjöll volcanoes can be accomplished in several ways. Some may do it the hard way, on foot or by mountain bike, whereas others can enjoy the luxury of a four-wheel-drive vehicle. The third option is to take the bus from Reykjhlíð at Mývatn, which runs daily over the summer time. The journey to Landmannalaugar and Veiðivötn can also be conducted in similar fashion. There are scheduled bus tours to the area from Reykjavík, Selfoss and Hella.

Askja–Kverkfjöll

On the way

Travelling from Mývatn towards the Hrossaborg tuff cone (Fig. 7.1), where the turn-off to Askja is located, it is worth stopping at the top of the *Námafjall* [65.6391, −16.8204] Ridge. There a magnificent view opens up towards the southeast over the northern part of *Ódáðahraun* [65.1804, −16.7715], the largest wasteland of lava in Iceland. Above the endless lava fields rise many móberg ridges and table mountains, and the most prominent of these are the table mountains *Búrfell* [65.5521, −16.6504] and *Bláfjall* [65.4422, −16,8439]. The oldest lavas are more than 3000 years old and most of them are derived from the lava shield *Ketildyngja* [65.4465, −16.6558]. The Búrfell lava immediately east of Námafjall was erupted by the cone row *Kræðuborgir* [65.6223, −16.5682] and is just over 3000 years old. The youngest lava is Nýjahraun ('new lava'). It originated at the *Sveinaborgir* [65.6251, −16.4039] cone row in 1875, a 25 km-long volcanic fissure, known locally as *Sveinagjá* [65.4410, −16.4822]. The eruption of this lava occurred on 18 February 1875 during the Askja Fires, a volcano-tectonic episode within the Askja volcanic system that began in 1874 and lasted until 1876. Here, out on the Askja fissure swarm, these fires were characterized by effusion of basaltic magma, whereas at the Askja central volcano the activity was very different. There, the main phase of the activity was a Plinian eruption (see p. 225–228).

On the east side of Námafjall the highway traverses a gently undulating landscape, the product of north–south-trending graben formed by faulting in association with activity on the Krafla, Fremri-Námur and Askja fissure swarms. The characteristic feature of these graben systems is their limited width in relation to their length. For example, the previously mentioned 30 km-long Sveinar graben (p. 174) is nowhere more than 0.6 km wide. The vertical displacement on the graben faults is 2–20 m. Many of the graben systems contain one or more cone rows, and the age relations between the faults and the cones show that in many instances the graben were formed earlier than the cone rows.

About 3 km before the large bridge across Jökulsá á Fjöllum, turn onto the track leading towards the interior by the volcanic cone *Hrossaborg* [65.6129, −16.2616] that is situated 1.2 km to the south of Highway 1 (see Fig. 7.1). This arena-like enclosure was part of the stage set in the 2013 feature film Oblivion. Road signs should indicate *Herðubreiðarlindir* [65.1963, −16.2238] and *Askja* [65.0348, −16.7592]. We are now well within the domain of the Askja volcanic system.

The Askja volcanic system – geology and history

The 200 km-long and less than 20 km-wide Askja volcanic system is the longest of its kind in Iceland (see Fig. 1.5). The central volcano, Askja (1550 m high), is situated in the southern sector of the system. The volcano has been active for several hundred thousand years. It rises to more than 800 m above its surroundings and is capped by an 8 km-wide composite caldera (see Table 1.4). The northern limb of the Askja fissure swarm is about 150 km long and contains many Holocene basalt volcanoes (i.e. cone rows and associated lava flows), including the previously mentioned Sveinagjá fissure and the Hrossaborg tuff cone. Farther south are the table mountain *Herðubreið* [65.1753, −16.3464] and the móberg ridge *Herðubreiðartögl* [65.1035, −16.3544], formed 258 000 years ago, which are surrounded by the lava shields *Kollóttadyngja* [65.2186, −16.5531], *Flatadyngja* [65.1345, −16.4666] and *Svartadyngja* [65.1075, −16.5278]. The southern limb of the fissure swarm is much shorter at about 30 km long. The main Holocene basalt eruptions are those of *Gígöldur* [64.8817, −16.9404], *Þorvaldshraun* [64.9686, −16.7140], and the three *Holuhraun* [64.8426, −16.8452] events, which took place in 1797, 1867 and 2014–15. Þorvaldshraun and the three Holuhraun events took place in historical times.

Locality 10.1   Hrossaborg – the horse pen

Hrossaborg is a tuff cone formed by an explosive eruption on the Askja volcanic system about 11000 years ago. It has a volume of 0.1 km³ and is situated about 75 km north of the Askja central volcano (Fig. 7.1). Its arena-like shape (Fig. 10.1) is attributed to partial collapse of the outer crater walls during the eruption, which was driven by erosion induced by flood-waters from the Jökulsá River at the time of the eruption. As a consequence the cone features excellent outcrops for exploring the volcanic architecture of a stratified tuff cone, exposing a range of hydromagmatic fall, surge and flow deposits, as well as impressive ballistic impact structures.

Interestingly, the Hrossaborg eruption began with lava fountain (i.e. dry) activity but it was short lived. Flooding by the Jökulsá River, induced by a

Figure 10.1 Hrossaborg tuff cone and its two stratigraphic sequences. (a) View of the Hrossaborg tuff cone from the southwest showing the extent of the lower and upper stratigraphic sequences in outcrop. (b) The two stratigraphic sequences at the far left sector of the outcrop in (a). (c) A close up of the lower sequence exemplifying the bedding and the two lithologies that make up the unit: ar, ash-rich lenses; lr, lapilli-rich lenses). (d) close-up of the upper sequence showing its overall appearance and highlighting the stratigraphic arrangement of the three key lithologies: cb, planar to cross-bedded tuff; m, massive lapilli tuff and al, the capping accretionary lapilli-bearing tuff. (e) ballistic lava blocks sitting in impact craters at the top of the Hrossaborg tuff cone succession. (f) Magnification of the sequence shown in (c) showing the abundance of armoured (ash-coated) basaltic pumice clasts within the sequence. (g) Magnification of the top al unit in (d); the pale bean-like features are the accretionary lapilli (= hemispherical aggregates of ash).

graben-like subsidence along the vent site, made the stage more ferocious and changed the event into an explosive hydromagmatic (i.e. wet) eruption. The Hrossaborg succession is divided into the lower and upper sequence (Fig. 10.1a), each reflecting distinct styles of explosive activity and modes of deposition for the erupted material.

The lower sequence is more than 40 m thick (Fig. 10.1a) and is typified by diffuse bedding on a decimetre scale, where indistinctly planar to cross-bedded tuff lenses alternate with basalt pumice bearing lapilli tuff or lapilli lenses (Figs 10.1c and f). This sequence is formed during sustained explosive activity typified by near-continuous explosions over a period of days, resulting in effectively simultaneous but sequential deposition of tephra from a series of laterally moving base surges and fallout from an elevated eruption plume. The tuff layers represent surge-dominated deposition mode, whereas the lapilli lenses are dominated by the air-fall (rain-down) mode. Syn-eruption erosion by the laterally moving surges explains the lensoidal form of the lapilli units.

The upper sequence, which is about 10–12 m thick, has a distinctly more chaotic look because it has been deformed by multitudes of 10–100 cm-large ballistic blocks that repeatedly impacted the otherwise distinctly bedded tephra deposits at the time of its construction (Figs 10.1b and d). The bedded tephra is typified by a sequence of planar to cross-bedded tuff that is overlain by a massive tuff or lapilli tuff containing fairly evenly distributed centimetre- to decimetre-large basalt lava blocks (grey), which in turn is capped by a less than 10 cm thick tuff unit containing abundance of accretionary lapilli (Figs 10.1 d and g). The upper sequence was constructed by a series of discrete explosions, where each event began with a surge moving laterally from the base of the rapidly rising and tephra-laden eruption column, forming the planar to cross-bedded tuff unit. This was followed by a partial but en-masse collapse of the rising eruption column, producing pyroclastic flows that laid down the massive lava block-bearing units. During the waning stage of each explosion ash aggregates were formed in the then ash-laden eruption plume, which, upon deposition, formed the accretionary lapilli-bearing units. At various stages, the tephra that was laid down was impacted by even larger ballistic lava blocks. This sequence of events was repeated by each explosive event, thus constructing the upper sequence as we see it today. The original outer cone wall surface, where preserved in its pristine state, comprises a 0.5–2.0 m-thick basalt block breccia that was formed, as indicated by the numerous ballistic impact craters (Fig. 10.1e), by phreatic explosions during the waning stages of the eruption.

On the way

Beyond Hrossaborg the geology is rather monotonous all the way to Herðubreiðarlindir, and the road follows an old sandur plain formed by torrential jökulhlaups induced by volcanic activity at or near the central vol-canoes of Kverkfjöll and Bárðarbunga, some 100 km to the south (see Fig. 1.5). At Grímstaðanúpur the track lies across pāhoehoe lava from the lava shield Kerlingardyngja, which was formed by a prolonged effusive eruption in the early Holocene. This is a good place to inspect pāhoehoe lava with well-developed tumuli and ropy lava surfaces.

Locality 10.2
Herðubreiðalindir – an oasis in rugged lava landscape

On the way to *Herðubreiðarlindir* [65.1963, −16.2238] we first cross two rivers, *Grafarlandaá* [65.3366, −16.0701] and *Lindá* [65.2296, −16.1851]. Both are spring fed with clear mountain water. The springs are located at the margins of the pristine-looking lava flow fields visible to the south and west of the track. The younger one, which the track crosses over shortly before the crossing over Lindá, is the late Holocene – possibly <2000 years old – lava shield Flatadyngja ('flat shield'). The older lava to the west belongs to the Kollóttadyngja lava shield that was formed in a prolonged eruption (lasting for decades?) more than 4500 years ago. These lavas are a natural water filter; rainwater seeps right into the porous lava, and impurities are filtered from it as the water is transported to the margins of the flow field as ground-water. Thus, when the water emerges as springs it is not only crystal clear but also refreshingly chilled.

Herðubreiðarlindir, or 'the oasis of Herðubreið', sits under the majestic tuya, Herðubreið (1682 m; Fig. 10.2a). It is an excellent place to spend the night, using the evening to explore the geology in the area. The lower part of Herðubreið is mainly built of subglacial volcanics, and the mountain is capped by a 300 m-thick lava shield. The transition from subglacial to sub-aerial volcanics occurs at about 1350 m, indicating that the ice sheet was at least 800 m thick when the subglacial eruption that sculptured the mountain into its present shape took place.

On the way (road to Drekagil at Askja)

The first part of the route takes you through pāhoehoe flows of the Flatadyngja lava shield, and then through rougher basalt lava flow fields that are derived from the Askja volcano. To the west of the track is the 258 000 year old móberg ridge Herðubreiðartögl ('tail of Herðubreið'). Here you may notice a patchy cover of pale pumice on the lava flows. These patches are the remnants of the rhyolite fall deposit from the 1875 Plinian eruption at Askja (see p. 225–228). Progressing south towards Askja volcano, the pumice fall covers a larger and larger area, until it completely blankets the surface at the southern end of Herðubreiðartögl and thickens rapidly towards the volcano.

The southern end of Herðubreiðartögl is a good stop to take a brief look at the surroundings. The Kverkfjöll central volcano rises above the horizon in the south, and the vast ice expanse in the background is the mighty

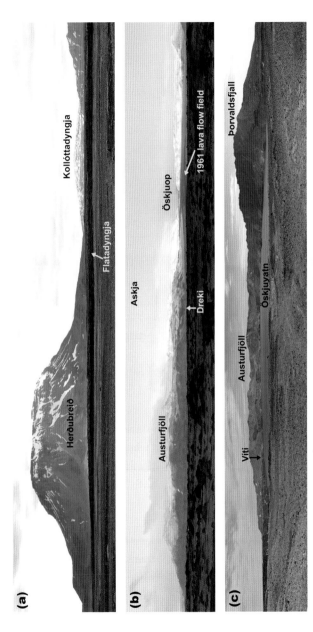

Figure 10.2 (a) The majestic table mountain Herðubreið viewed from the northeast just before arriving at Herðubreiðarlindir (just off the photograph to the left). In the foreground is the fluvial plain of Jökulsá á Fjöllum, then the lavas of Flatadyngja. Kollóttadyngja rises from the horizon to the right. (b) Askja volcano as seen from the east at the south end of Herðubreiðartögl. The whole mountain range from far left to far right of the photograph is the volcano. The mountains to the left of centre are the Austurfjöll sector of the volcano with the Dreki hut at their northern end. Öskjuop ("Askja entrench") is the passageway into the Askja caldera. The 1961 lava flow field (black) extends out through Öskjuop. In the foreground are older Askja lavas, now partly covered by the 1785 Askja Plinian fall deposit (whitish yellow). (c) Öskjuvatn caldera. View is to the east towards Austurfjöll. The Víti phreatic crater is located on the north bank of the caldera lake as indicated. In the foreground is the pumice fall from the March 2–29 1875 event.

Vatnajökull ice sheet. Looking to the west, the view opens to the Askja volcano (Fig. 10.2b) and to the east the pillow lava ridges Miðfell and Upptyppingar rise above the lava plain. Still farther east is the móberg ridge Fagradalsfjall and, if weather permits, the ice-capped Snæfell stratovolcano is visible. Here the 1875 pumice fall rests on the lavas of Svartadyngja ('black lava shield') which were emplaced after the H5 event at the Hekla volcanic system about 7000 years ago but before the H4 event from the same system 4200 years ago. A little farther south it rests on more rugged lavas of similar age, but originating from vents on the eastern flanks of the Askja volcano. In front of Drekagil ('dragon's gorge') on the east flank of Askja is the Dreki ('dragon') mountain hut (Fig. 10.2b). Drekagil features excellent outcrops of products of subglacial eruption such as pillow lavas, kubbaberg breccia and móberg tuff.

Locality 10.3 The Askja volcano

From Dreki, the track meanders through a pristine basalt-lava flow field, gradually climbing to Öskjuop, a 1 km-wide pass through the eastern wall of the Askja caldera, situated at an altitude of 1100 m. A pitch-black lava stands out against the pale pumice-fall deposit of 1875 that covers most of the older lava formations at Öskjuop (Fig. 10.2b). This is the youngest lava flow at Askja and was formed in an effusive eruption in 1961.

The 1961 eruption began on 26 October on a 0.7 km-long fissure running east–west along the junction between Öskjuop and the eastern boundary fault of the main Askja caldera. In the beginning, the lava was spurting out as 500 m-high fountains, which within ten hours had produced a 7.5 km-long fountain-fed 'ā'a lava, which covered an area of 6 km². This was followed by an eruption of smooth-surface pāhoehoe lava, which enlarged the lava flowfield to 11 km2 by the time the eruption ended five weeks later. The spatter and scoria cones that delineate the 1961 fissure are the ones that rise over the parking lot at Öskjuop.

From the parking lot there is about a 20-minute walk to the explosion crater Víti and Öskjuvatn ('Askja lake'), which occupies a small collapse caldera formed in the Plinian eruption in 1875 (Fig. 10.2c). The Öskjuvatn caldera is nested within the much larger main Askja caldera, which measures about 8 km across. The steep-sided mountains and cliffs that surround this area on all sides are the rim of the main caldera.

The 1874–1930 Askja Fires

The 1874–1930 volcanotectonic episode, which includes the explosive Plinian eruption on 28–29 March 1875, is one of the best-recorded historical events on the Askja volcanic system. This episode involved rifting along the Askja volcanic system, with eruptions at the central volcano and out on the fissure swarm. Therefore, this episode is collectively referred to as the Askja Fires.

Observations from the Mývatn area, some 100 km to the north, indicate that precursory activity began as early as February 1874, when dense columns of steam were seen rising from Askja. In the last two weeks of December 1874, strong and frequent earthquakes were felt in northern Iceland, and columns of smoke and fires were seen rising from Askja on 1 and 3 January 1875. Minor ashfall occurred in the districts south of Öxarfjörður in North Iceland. At this time there may have also have been an eruption just north of Askja.

A team of farmers from Mývatn visited the Askja volcano on 15 February 1875. They noticed that the southeast corner of the main caldera (i.e. the area of present-day Öskjuvatn) had subsided about 10 m and that a crater spewing mud and debris had formed just west of the subsidence. Three days later a basaltic fissure eruption began on the 30 km-long Sveinagjá fissure some 40 km north of the Askja volcano. This fissure erupted intermittently for several months and produced 0.2–0.3 km³ of lava.

An explosive Plinian eruption began within the Askja caldera in the early hours of 29 March. The eruption plume was dispersed eastwards by strong winds and by 3.30 a.m. tephra fall was reported in Jökuldalur about 70 km to the east of Askja. The tephra was greyish in colour and had a fine- grain texture like ash; it was also very wet and sticky. This first phase of the explosive eruption lasted for an hour. At 5.30 a.m. the tephra fall had dwindled and the air had become somewhat clearer. But this was just the calm before the storm, because shortly afterwards the main Plinian eruption began. Light-brown pumice of ever-increasing grain size rained down until noon. The plume was carried onwards across the Norwegian Sea, and tephra fall was reported in Scandinavia (Fig. 10.3). Shortly after noon on 29 March the tephra fall had stopped over populated areas in Iceland, but the eruption raged on at Askja until the following day. Soon afterwards, on the north margins of the lake the maar-like explosion *Víti* [65.0462, −16.7263] was formed in a shortlived hydromagmatic eruption. Today, it contains fumaroles and a small lake with water warm enough for bathing.

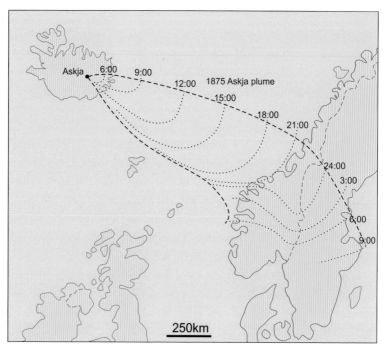

Figure 10.3 Dispersal times for the Askja eruption plume. Position of the front of the plume at specific times during the course of the days 28 and 29 March 1875 is indicated by the dotted lines.

Eruptive activity resumed in the Askja system between 1920 and 1930, with five separate eruptions occurring on ring fractures around the Öskjuvatn caldera and an eruption on a 6 km-long fissure on the southern flanks of the volcano that produced the Þorvaldshraun lavas.

In 1876 the Danish geologist Johnstrup visited the Askja volcano, together with his fellow countryman, the surveyor Caroc. They observed that a roughly circular area of subsidence had formed in the southeastern corner of the Askja caldera. It measured 4580 m by 2500 m and the deepest part was 234 m below the floor of the main Askja caldera (see also first panel on Fig. 10.4). This subsidence was bordered by many concentric faults, and at its bottom was a small lake. Johnstrup and his companion were observing a caldera collapse. The roof of the crustal chamber beneath the Askja volcano had collapsed into the void left behind by the magma withdrawn from the holding chamber during the 1875 Plinian eruption. The Öskjuvatn caldera was born, the first of its kind in Iceland in historical times. The caldera continued to grow in size for a significant amount of time after the March 1875 events, and it took more than five decades for it to

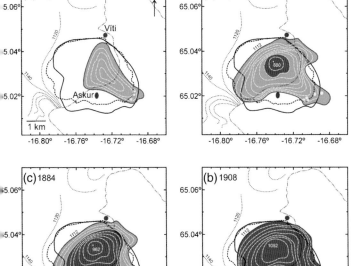

Figure 10.4a Five stages in the evolution of the Öskjuvatn caldera between 1875 and 1932. The outline of Öskjuvatn caldera in 1932 is shown by a solid blue line and the outline of the modern day Öskjuvatn is shown by the blue dashed line. The western edge of Austurfjöll prior to 1875 is indicated by the orange dashed line. Orange shaded areas represent areas of subsidence; blue shaded areas the lake growth.

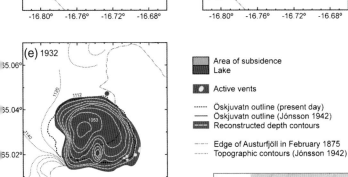

	Area of subsidence
	Lake
	Active vents

- - - - - Öskjuvatn outline (present day)
———— Öskjuvatn outline (Jónsson 1942)
▓▓▓▓ Reconstructed depth contours

—·—·— Edge of Austurfjöll in February 1875
·········· Topographic contours (Jónsson 1942)

Figure 10.4b Bathymetry of the Öskjuvatn caldera with a colour coded depth scale (see key). Also shown are the 2014 subaerial rockslide scar (solid black line) and sub-aqueous rockslide deposit (broken black line) and location of active and inactive underwater geothermal vents (stars). The rockslide originated from the highly unstable south caldera wall which rises >450 m above the lake surface.

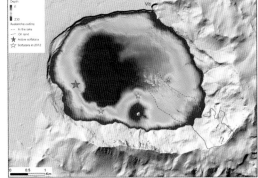

reach its full size (Fig. 10.4a). The lake Öskjuvatn did not reach its present water level until 1907; thus, it took 32 years to fill the depression with water.

The present level of the lake is 50 m below the caldera floor, and recent bathometric soundings reveal a maximum water depth of 224 m. Areas of vigorous geothermal activity are found along the eastern and southern margins of the lake. Its area is 10.7 km^2 and the volume 1.2 km^3. The total volume of the subsidence is about 2 km^3, which is almost 10 times the volume of magma erupted in the 1874–1930 volcanotectonic episode. This implies that a considerable amount of magma involved in this episode is stored as intrusions within the volcano or the fissure swarm. Askja is again showing signs of unrest. In 2012, the Öskjuvatn lake was ice-free in March (i.e. middle of winter), which is very unusual and indicates increased geothermal activity in the fumarolic fields within the lake. Just before midnight on 21 July 2014 a major rockslide came from the unstable southeastern caldera wall, i.e. in the region of Suðurbotnar, and cascaded into Lake Öskjuvatn, producing a tsunami with an amplitude of >30 m (Fig. 10.4b). It is possible that this landslide is a consequence of the heightened geothermal activity within the volcano.

Locality 10.4   The 2014–15 eruption at Holuhraun

The 2014–15 Holuhraun eruption was an archetypical Icelandic eruption, an eruption from a 1.8 km-long fissure that lasted from 31 August 2014 to 27 February 2015. It was the largest event of its kind in Iceland for 236 years, or since the Laki flood lava eruption in 1783–4. The 2014–15 eruption discharged ~1.2 km^3 of magma at an average rate of 77 m^3/s, producing a lava flow field covering ~84 km^2 of the relatively flat flood plain in front of the outlet glacier Dyngjujökull (Fig. 10.5a and b). The activity at the vents featured up to 100 m high lava fountains that supported a 1–4 km-high, gas-charged eruption plume that produced significant volcanic (sulphuric) pollution across Iceland (Fig. 10.6a and b). From 31 August to mid-October the lava was discharged from the vent at rates of ~570 to 100 m^3/s via open channels, forming four slabby to rubbly pāhoehoe to 'ā'a lava flows (Fig. 10.6c, e, f) emplaced side by side and up to 18 km long (Fig. 10.5b). Early on, the lava fountains produced abundance of Pele's hair (Fig. 10.6g).

In the period mid-October to end November the discharge dropped to ~100–50 m^3/s. Subsequently, lava fountaining stopped, and the vent activity was typified by periodic bursting of very large bubbles, up to 50 m in diameter (Fig. 10.6d). At this time, the lava was delivered via a secondary lava pond

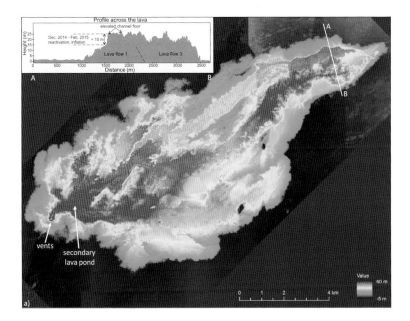

Figure 10.5 (a) A map showing the thickness of the 2014–15 lava flow field at Holuhraun. Maximum thickness is just over 60 m (dark red areas; see scale bottom right). Also indicated are the vents, the secondary lava pond and lava flows 1 and 3. Inset illustrates the effects of the inflation caused by the December 2014 to February 2015 reactivation of flows 1 and 3. (b) Map showing the extent and timing (see key) for individual lava flows of the 2014–15 flow field. The numbers indicate the first five flows, where the first flow is the only one extending the entire length of the flow field and subsequent flows are largely emplaced sequentially on its south side. Figures in the inset (top left) give the length (L), area (A) and volume (V) of the flow field.

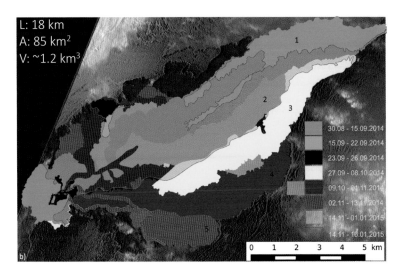

Figure 10.6 Photographs from the 31 August 2014 to 27 February 2015 eruption at Holuhraun. Shooting day is given in the lower right corner of each photograph. (**a**) Sulphur plumes rising from the fountaining fissure in the early days of the eruption. Lava fountains up to 100 m high. View is to east. (**b**) Vent activity largely concentrated on the central part of the 1800-m-long fissure in the 5th week of the eruption. The tallest fountain rises more the 100 m into the atmosphere. View is to east. (**c**) Lava river extending from the vents to the active flow front in mid-October. Note, no lava fountains at the vents, and the front (lower right) of the lava is covered by crustal plates. View is to west. (**d**) View of the lava pond in the vents in mid-November. The large bubble (arrow) is about 40 m in diameter and the rim to rim distance is ~100 m. View is to north-northeast. (**e**) Sheet lobe of slabby pāhoehoe forming in the first day of the eruption, with up to one-metre-wide slabs (s). Arrow points to mounds of gravely sand pushed up by the moving lava. (**f**) Flow margins of a rubbly pāhoehoe lobe on the north side of the lava flow field about 7 km from the vents. The lava surface features 1–3 m plates formed by break-up of pāhoehoe crust (black arrow) sitting in a rubble of equant decimetre blocks (white arrow). The incandescent lava represents a small lava breakout from the base of the rubbly pāhoehoe lobe, forming a new lobe of ´ā´a. (**g**) Bundle of Pele´s hair formed by the lava fountains in the first week of eruption. (**h**) Steam plumes rising from the distal part of the lava flow field, which by mid-November was inactive. Water from the branch of the river Jökulsá á Fjöllum that followed the southern (left; arrows) margins of the flow field perco-lated under and interacted with the hot lava interior to produce the plumes. The large plume in the far distance is rising from the vents 17 km away. The dark and snow-free area is the 2014–15 lava flow field. View is to the west.

through a partly concealed transport system to form the four remaining flows of the lava flow field (Fig. 10.5a). For the remainder of the eruption the discharge was <50 m3/s and the lava was delivered to active flow fronts via internal pathways (i.e. lava tubes), which also reactivated two of the earlier-formed lava channels, resulting in elevation via inflation of the channel floor to form lava rise plateaus that now rise more than 10 m above the surrounding lava (Fig. 10.5a, inset). At this time, the internal pathways (i.e. lava tubes) fed surface breakouts of slabby to spiny pāhoehoe (e.g. Fig. 10.6e), which resurfaced about 19 km^2 of the flow field.

Locality 10.5   The Kverkfjöll volcanic system

The Kverkfjöll volcanic system (see Fig. 1.5) is about 125 km long. The main volcanic structures are the *Kverkfjöll* [64.6874, −16.6027] central volcano (1929 m high) with its glacier-filled twin calderas and the mountain range *Kverkhnjúkar* [64.7682, −16.4936], which constitute the most volcanically active part of the fissure swarm. One of the outstanding features of Kverkhnjúkar is a series of volcanic ridges, 10–30 km-long, that extends northwards from the central volcano and rise 200–300 m above the surrounding plains (Fig. 10.7). These ridges are composed almost entirely of pillow lava, which implies that they were formed by effusive fissure eruptions under a substantial thickness of glacial ice. Tuffs formed by explosive subglacial eruptions occur only on peaks that rise to more than 1000 m above sea level, showing that the Weichselian glacier (the last glacial period; see Chapter 1) was at least 400 m thick when these subglacial eruptions took place.

The valleys between the ridges are floored by many basalt-lava flows formed by subaerial fissure eruptions over the past 11 500 years. Some of

Figure 10.7 Kverkhnjúkar pillow ridges, view towards the north. Brúarjökull to left.

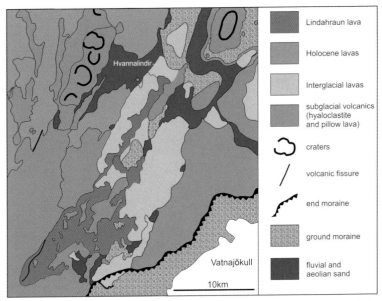

Figure 10.8 Geology of the area around Hvannalindir, showing the distribution of the Lindahraun lava and other geological formations.

Figure 10.9 Tephra layers in a soil profile at Hvannalindir. The numbers are years of eruption, with the source volcanoes named.

these eruptions must have featured spectacular lava fountains, because they produced extensive aprons of fountain-fed lava mantling every hill and peak up to 1.5 km from the source vents.

Hvannalindir – a young oasis in a highland desert
Hvannalindir [64.8692, −16.3267] is a small oasis in the otherwise barren highlands north of the Kverkfjöll volcano. It is situated in front of the *Lindahraun* [64.8513, −16.3195] lava and owes its existence to the freshwater stream Lindá that emerges as cold-water springs at the lava front on the south side of Hvannalindir (Fig. 10.8).

The sandy soil cover at Hvannalindir is up to 2 m thick and contains several tephra layers and three thick units of aeolian sand. The oldest tephra layer in the soils of Hvannalindir indicates that the oasis began to form about 1300 years ago, two centuries before the human colonization of Iceland (Fig. 10.9). More importantly, this also gives the minimum age (AD 700) for the Lindahraun lava, which is the youngest volcanic formation on the Kverkfjöll fissure swarm. These results show that no volcanic eruption has occurred on the Kverkfjöll volcanic system outside of the central volcano in historical times (i.e. the past 1200 years), which is surprising because the Kverkfjöll system is situated above the centre of the mantle plume.

The youngest tephra layer in the soil at Hvannalindir is the rhyolite tephra fall from the 1362 eruption at Öræfajökull. This shows that, until the fourteenth century, the climate was favourable and the Hvannalindir oasis enjoyed stable, and perhaps lush, vegetation cover. The aeolian sand unit towards the top of the soil cover shows that some time after the fourteenth century the conditions changed for the worse. The climate deteriorated as the Little Ice Age set in and, because of stronger winds, the oasis was inundated by sand instead of silty loess; thus, the vegetation deteriorated. The oasis has not yet recovered from the setback of the Little Ice Age.

In the eighteenth, century Hvannalindir was the refuge of Iceland's most famous outlaw, Fjalla-Eyvindur ('Eyvindur of the mountains'). In those days it was a perfect hiding place because it was secluded and protected by large glacial rivers on three sides and on the fourth by the Vatnajökull glacier. Fjalla-Eyvindur lived here for almost a decade. The ruins of his home can be seen near the edge of the Lindahraun lava in the southeast corner of the oasis. Here Fjalla-Eyvindur built a rather elaborate home out of rocks and turf, containing separate sleeping quarters, a kitchen, and a toilet that included an automatic flushing system (the river), which would have been a luxury in those days.

Landmannalaugar–Veiðivötn

This excursion is centred on the popular Torfajökull–Veiðivötn region, which features some of the most colourful rock formations in Iceland and spectacular rows of explosion cones and craters.

Locality 10.6   The Torfajökull volcano – complex accumulation of rhyolite lavas and domes

The *Torfajökull* [63.9014, −19.0220] central volcano in South Iceland is the largest caldera volcano in Iceland, with a basal diameter of 30 km, and is crowned by an 18 km-wide caldera (see Table 1.4), which hosts the largest geothermal area in Iceland. This rugged mountain region is by far the largest area of rhyolite in the country, where colourful subglacial and subaerial volcanics cover about 350 km^2 (Fig. 10.10). The Torfajökull volcano is an anomaly in the predominantly basaltic volcanic succession of Iceland, because the ratio of rhyolites to basalt is about four to one, which is in stark contrast to the 1:5 ratio observed for most of the other central volcanoes. This anomaly reflects the setting of the volcano, which is situated at the boundary where the East Volcanic Zone changes from being a simple rift zone, characterized by spreading away from the rift axis, to a forward-propagating rift zone. The Torfajökull volcano exhibits distinct polarity in terms of the composition of the erupted magmas. It produces mostly rhyolites and basalts, and rocks of intermediate (andesitic) compositions are present only in minor amounts. This compositional gap, often referred to as the Daly gap, implies that the basalt and rhyolite magmas are not derived from the same source. The basalt magmas originate from a deeper source within the mantle plume, whereas the rhyolite magmas are produced by partial melting of basaltic crust at shallow depths (3–10 km).

The oldest stratigraphical units at the Torfajökull volcano date back to the most recent interglacial period (100 000 years ago). These are thick rhyolitic lavas and welded airfall tuffs. This was followed by large ring-fracture eruptions forming a series of subglacial rhyolite lava domes (e.g. *Bláhnjúkur* [108 ka; 63.9771, −19.0675], *Kirkjufell* [273 ka; 63.9841, −18.9443], *Torfajökull* [63.9014, −19.0220], *Laufafell* [373 ka; 63.9154, −19.3414] and *Rauðufossafjöll* [67-77 ka; 63.9961, −19.3705]) that encircle the entire caldera (Fig. 10.10).

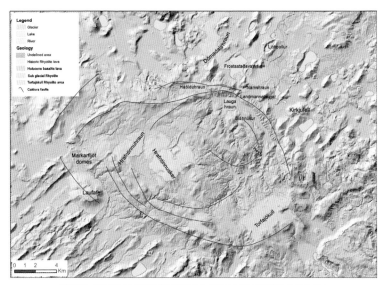

Figure 10.10 Simplified geological map of the Torfajökull volcano.

Some of these lava domes grew large enough to penetrate the ice sheet and form silicic table mountains.

In postglacial times the Torfajökull volcano has produced several relatively thin (< 50 m) obsidian lava flows that appear as dribbles of black icing on the brightly coloured landscape. These obsidian lavas are *Dómadalshraun* [64.0296, −19.1009], *Hrafntinnuhraun* [63.9526, −19.2744] ('obsidian lava'), *Námshraun* [63.9998, −19.0483] and *Laugahraun* [63.9881, −19.0735] (Fig. 10.10). Dómadalshraun is the oldest, formed in AD 150, then Hrafntinnuhraun (AD 870), followed by the eruption of Námshraun and Laugahraun in AD 1477. All of these lavas were erupted in association with major basalt-fissure eruptions on the adjacent Veiðivötn volcanic system (see locality 10.7) and they also contain an abundance of mafic-melt inclusions, indicating that rhyolite residing in a shallow crustal magma chamber was reheated and mobilized by injection of hotter basalt magma. The youngest rhyolite formations at Torfajökull are erupted on northeast–southwest orientated fissures, implying that at present the activity at the volcano is mainly controlled by normal rift-zone tectonics.

Locality 10.7
Veiðivötn and Vatnaöldur fissures, cones and maars
The Veiðivötn volcanic system (see Fig. 1.5) is about 150 km long and takes its name from the many lakes and ponds of *Veiðivötn* [64.1419, −18.7662] ('lakes

235

Table 10.1 List of Holocene and historical basaltic flood lava and fissure eruptions in the Veiðivötn area. Ages are listed as years before present (BP) using 2000 as the reference year. Number in parentheses indicate calendar years (AD)

Lava flow or Eruption	Age (B.P.)	Length (km)	Area (km²)	Volume (km³)
The Great Þjórsárhraun lava	8600	130	950	24
Tungnaárhraun (THc) lava	6800	55	120	1.4
Tungnaárhraun (THd) lava	6600	80	270	3.8
Tungnaárhraun (THe) lava	6400	75	260	1.0
Kvíslahraun lava	6200	65	200	3.4
Hnubbahraun lava	6000	15	25	0.4
Búrfellshraun –Þjórsárdalshraun lava	3000	70	350	6.5
Tjörvahraun – Hnausahraun lava	1850 (150AD)	20	55	0.8
Vatnaöldur (explosive)	1130 (870AD)	–	–	3.3
Veiðivötn (explosive)	520 (1480AD)	–	–	3.5
Tröllahraun lava	138 (1862AD)	25	28	0.3

of good fishing'), which are known for their trout. The most prominent volcanic landforms in the system are the glacier-covered Bárðarbunga central volcano at the western edge of Vatnajökull icecap and the volcanic fissures that transect the Veiðivötn basin from northeast to southwest. The Veiðivötn system is one of the most productive volcanic areas in Iceland in Holocene and historical times. In the past 800 years the Bárðarbunga volcano has erupted at least 17 times and at least 12 major basalt-fissure eruptions have occurred in the Veiðivötn basin in the past 10 000 years (Table 10.1). The lavas produced by the Holocene flood-lava eruptions, which include the Great Þjórsá and the Þjórsárdalur flows (see pp. 77), generally followed the course of the Tungnaá and Þjórsá rivers as they found their way towards and onto the Southern Lowlands). This is a major concern, as lava flows from future eruptions may follow the same general path and destroy the hydroelectric powerplants on these rivers at and above Mt Búrfell.

Today, most of the Holocene fissures and craters are hidden from view by the many lakes situated in tuff cones, tuff rings and maar craters that floor the Veiðivötn basin. Most of these cones and craters were formed in three fissure eruptions in AD 150, 870 and 1477. The first of these eruptions occurred on an 8 km-long volcanic fissure at the southwestern end of the basin (Fig. 10.11). The southern end of the fissure extended into the Torfajökull volcano and produced the silicic Dómadalshraun, whereas the northern part of the fissure erupted basaltic magma and formed two small basaltic lava flows known as

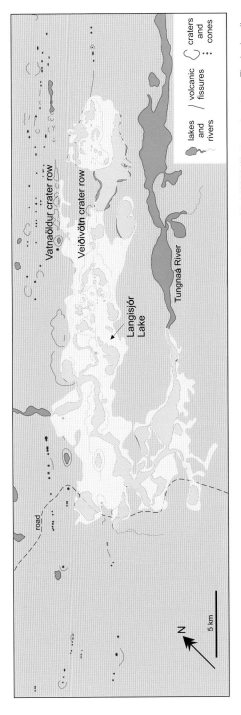

Figure 10.11 The main features and position of the AD 150 Hnausahraun–Tjörvahraun, AD 870 Vatnaöldur, and AD 1480 Veiðivötn vent systems. The broken line shows the approximate outlines of the Langisjór lake formed as a consequence of the eruption in AD 150.

Hnausahraun [64.0682, −19.0452] and *Tjörvahraun* [64.1356, −19.1510]. These lavas flowed into and dammed up the Tungnaá River, which consequently flooded the Veiðivötn area, forming a basin-wide lake in the process. Thus, when the 870 eruption took place and formed the *Vatna*öldur [64.1175, −18.9173] cone and crater row, the basin was submerged in water (Fig. 10.11). As the magma emerged from the new fissures, it encountered and interacted explosively with the lake water. As a result, the eruption was almost entirely explosive and it covered roughly half of Iceland with basaltic tephra. As in the previous eruption, the feeder dyke for the Vatnaöldur eruption cut through the Torfajökull volcano and thus initiated an eruption adjacent to *Hrafntinnusker* [63.9338, −19.1891], which produced rhyolite tephra, along with the obsidian lava of Hrafntinnuhraun. The Vatnaöldur eruption pre-dates the arrival of the first settlers in Iceland by a few years, and its tephra layer is also known as the Settlement layer, which is an important time marker in the soils of Iceland for both volcanological and archaeological studies. After the Vatnaöldur eruption, the lake was re-established, and thus the events of 870 were repeated when the 1480 eruption occurred and produced the 40 km-long Veiðivötn cone and crater row (Fig. 10.11). The Vatnaöldur and Veiðivötn eruptions produced more than 3.3 km^3 and 6.7 km^3 of basaltic tephra, respectively, which makes them the two largest explosive hydromagmatic basalt fissure eruptions in recent history.

Addendum

2021 Fagradalsfjall eruption

Geological setting

The Reykjanes Peninsula links the onshore Western Volcanic Zone and the South Iceland Seismic Zone of Iceland to the offshore Reykjanes Ridge (Fig. 1.2). It is taken to be a highly oblique spreading segment that is expressed by distinct surface fracture pattern that includes north–south

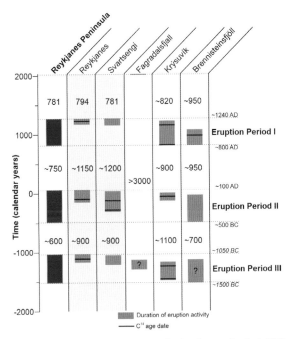

Figure A.1 Eruption periods on the Reykjanes Peninsula over the last 4000 years. The timescale is in calendar years, where AD (Anno Domini) is the common era age and BC (Before Christ) denotes ages predating the start of the common era. The orange bar indicates the extent of the eruption activity within individual systems, while the extent of each Eruption Period on the Reykjanes Peninsula is indicated by the red bars. The length of each repose period is indicated by the figures between the bars. Carbon-14 age determinations are indicated by black bars. The Reykjanes and Svartsengi volcanic systems are shaded grey to indicate that they are considered to represent a single system by some researchers. Modified from ISOR prepared figure from 2010.

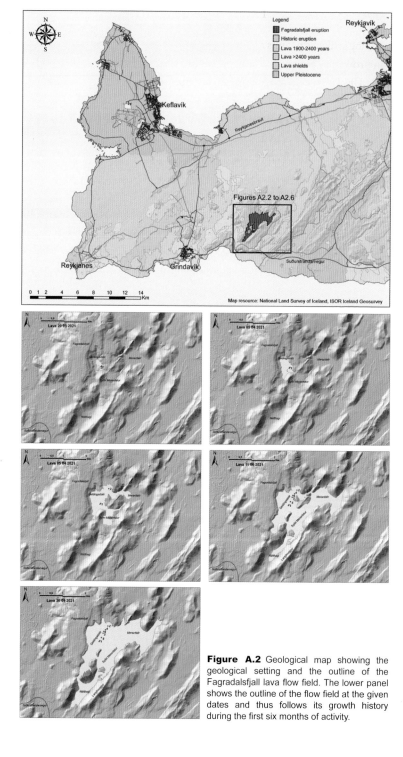

Figure A.2 Geological map showing the geological setting and the outline of the Fagradalsfjall lava flow field. The lower panel shows the outline of the flow field at the given dates and thus follows its growth history during the first six months of activity.

trending strike-slip faults and N21–40°E fractures trending as the volcanic cone-rows that define the active volcanic systems on the peninsula (Fig. 1.5). Until recently, Fagradalsfjall has not been included as one of the active systems, but as the current volcanic event on the peninsula demonstrates, it could represent a system of its own. In the last 4000 years the pattern of volcanic activity has been periodic, where 200–400 year-long eruption periods, when the whole of the peninsula was activated, are separated by 600–800 year-long periods of volcanic quiescence (Fig. A.1). The last eruption period culminated in 1240 CE or just over 780 years ago, which matches well with the length of the established repose period. Therefore, it is not unreasonable to assume that the ongoing eruption at Fagradalsfjall is signalling the onset of a new eruption period on the Reykjanes Peninsula.

The 2021 Fagradalsfjall eruption

After a repose of 781 years, eruptive activity resumed at the Reykjanes Peninsula, when at 20:30 (local time) on 19 March 2021, basaltic magma began to issue from a short linear vent system situated in dales of Geldingadalir in the Fagradalsfjall subglacial volcanic complex (Fig. A.2) and signalling the start of a new Eruption Period on the peninsula. The eruption was preceded by intensified seismic unrest in the Fagradalsfjall region that began on 24 February 2021. This seismic activity was initially linked to movements on the Reykjanes Peninsula plate boundary and the emplacement of a 5–7 km-long regional magma feeder dyke between Fagradalsfjall and Keilir (Fig. A.3).

The eruption began on a 180-m-long, north-northeast-trending fracture system situated above the inferred southern end of the regional feeder dyke and featured at least 10 erupting vents (Fig. A.2). This activity was quickly consolidated onto two vents, labelled 1a and 1b, which were the only active vents for two and a half weeks or until 5 April 2021. During this time, the venting of the magma was characterized by steady bubble-bursting to weakly fountaining activity accompanied by continuous outflow of lava that supplied the bulk of the lava initially emplaced within the confines of Geldingadalir. Between 5 and 13 April five new vents were formed along a 1-km-long lineament extending to the north from the original vents (Fig. A.2). All of the new vents featured eruptive behaviour similar to vents 1a and 1b. Towards the end of April activity ceased on all vents, except vent 5,

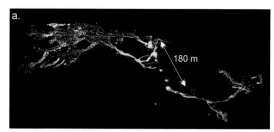

Figure A.3 Photographs of the activity at the Fagradalsfjall eruption. **(a)** Early morning on 20 March 2021 at 3:40 showing a series of active vents on the 180-m-long vent system and the lava streams produced by these vents in the first 8 hours of eruption.

(b) Vents 1a and 1b erupting on 4 April 2021 delivering pāhoehoe lava (dark grey) into Gelding-adalir via small lava pond (the light grey surfaced lava) that formed in front of the vents. The red-brown mounds in the pond are pieces broken off from the crater rims and then rafted down-stream.

(c) Lava fountaining through vent 5 and asso-ciated tephra fall on 9 May 2021.

(d) 2 to 5 m-thick active 'ā'a lava front in Meradalir advancing at the rate of 3–4 m per hour. The vegetation in front of the lava is scorched (black patches) from fires ignited by the lava.

(e) An open channel delivering lava (from right to left) from vent 5 into Syðri Meradalir on 27 April 2021. In the fore-ground a slabby to rubbly pāhoehoe (dark brown to black) is partly covered by pāhoehoe (light grey). In the far distance is rubbly pāhoehoe in Geld-ingadalir (right) and Syðri Meradalir (left).

which became the loci of activity. Since then, or for about four and a half months, vent 5 has episodically vented magma through lava fountaining of variable intensity and periodicity as well as steadily delivering lava to the flow field that completely or partly filled the vales of Geldingadalir, Meradalir, Syðri Meradalir and Nátthagi (Fig. A.2). The magma has been delivered from the vents via both (i) sealed internal lava pathways (= lava tubes) and (ii) open lava channels to produce the full spectrum of subaerial basaltic lava types, including pāhoehoe plus spiny, ropy, slabby and rubbly pāhoehoe, toothpaste lava and 'ā'a (Fig. A.3). The magma supply rate has been steady, in the range of $4–8m^3/s$ and as of mid-September the lava flow field covered 4.6 km^2 and its volume was just over 0.1 km^3. The magma erupted is olivine tholeiite with mean melt MgO content of 8–8.5 wt.% and contains 0–15% of plagioclase, olivine and rare clinopyroxene phenocrysts. The eruption has released between 2–4000 tons of SO_2 and 5–10 000 tons of CO_2 per day.

Glossary

agglutinate Pyroclastic rock made up of partly fused volcanic bombs and tephra.

caldera A large (>1 km in diameter) circular depression formed by down-sagging or collapse of a magma chamber roof as a result of withdrawal of magma from the chamber during eruption(s).

compound lava Lava flows that consist of multiple flow lobes in various lateral and vertical arrangements. Note that almost all lava flows consist of two or more flow lobes and thus in essence are compound. However, lavas that comprise long and thick flow lobes may superficially appear to consist of a single flow lobe (or flow unit). Such lavas are sometimes referred to as 'simple flows'.

corrie or **cirque** Landform of glaciated highland regions, created when there is enough snow accumulation at the head of an ordinary valley head to form a glacier large enough to begin to flow. The backwalls of the valley are cut back by plucking, whereas the valley floor is gouged out by the moving glacier. This erosion by the glacier forms a bowl-shape depression called a corrie or a cirque.

debris flow Rapid downslope movement of a mass of debris that exhibits the flow properties of a viscous fluid.

diamictite A non-genetic term used for chaotic and unstratified sediments that consist of clay- to boulder-size particles and are poorly sorted.

diatom A one-cell plant that has siliceous framework and grows in oceans and lakes.

diatreme A volcanic pipe (vent) formed into solid rock by a gaseous explosion and filled with volcanic breccia and lapilli.

dropstone Ice-rafted gravel to boulder size stones (i.e. rock fragments),which drop from melting iceberg into marine sediments.

eutaxitic texture See **fiammé**.

fall deposit A deposit of tephra formed by fall (or rain) out for pyroclasts from a volcanic plume raining. Also referred to as ash or tephra fall deposits.

fault A fracture along which the opposite sides have been displaced relative to each other.

faulting Deformation of crustal rocks by fracture.

felsic Refers to igneous rocks that are relatively rich in silica (+ natrium and potassium) and rich in the light-coloured mineral feldspar.

fiammé Deformed pumice clasts that occur in welded ignimbrites as pale brown to black glassy lenses with flame-like shapes. Fiammé exhibit bedding-parallel alignment of long dimensions, giving the rock a distinctive foliation referred to as **eutaxitic texture**.

fires In the old chronicles, Eldur ('fire') is often used to describe a volcanic eruption. It reflects direct experience of eruptions, which most commonly feature fire-fountain activity or, as perceived, columns of fires.

flow deposit A deposit of tephra formed by pyroclastic flow (or high concentration pyroclastic density current), which is a very fast-moving current of hot gas and pyroclasts formed in some volcanic eruptions.

fracture zone A linear feature in oceanic crust – often hundreds, even thousands of kilometres long – formed by actions that offset mid-ocean ridge axis segments.

graben A downthrown block between two parallel normal faults.

hornito A small spatter (i.e. driblet) cone formed around rootless vent within lava flows.

horst An elongate uplifted block bounded by faults on its long side.

hotspot Area of unusually productive volcanism generally considered to be the surface manifestation of a mantle plume.

hyaloclastite Volcaniclastic breccia formed by non-explosive fracturing and disintegration of quenched lavas. Fragmentation occurs in response to thermal stress built up during rapid cooling and stress imposed on chilled outer parts of lava flows by continued movement of the molten interior.

intraplate volcanic belts A volcanic zone located within a plate rather than at the boundary between two plates.

jökulhlaup An Icelandic term for a glacier outburst flood.

kubbaberg Cube-jointed basalt.

laccolith An intrusion injected between two layers of strata exhibiting a dome or mushroom-like form with a planar base.

lava fountain A red-glowing pillar or column of lava rising into the air from an active vent.

large igneous province A very large basaltic volcanic province characterized by emplacement of mainly mafic (i.e. Fe- and Mg-rich) extrusive and intrusive rocks, which originate via mantle processes other than normal seafloor spreading. For example, they are produced by mantle plumes.

lithification The process that changes a sediment into sedimentary rock.

lithology A description of the physical characteristics of a rock unit visible in an outcrop, hand or core samples or with low magnification microscopy, and includes properties such as colour, texture, grain size, or composition.

mafic An adjective describing a silicate mineral or rock that is rich in magnesium and iron and contains a relatively high abundance of dark-coloured minerals; the term is coined from the words '*ma*gnesium' and '*f*erric'.

magma Molten rock that solidifies upon cooling to form igneous rocks.

mantle plume Plume-shaped structure of hotter and less dense material within the Earth's mantle that is more buoyant than the surrounding mantle and thus rises towards the surface.

mid-ocean ridge A major elevated linear feature of the sea floor, marking the boundary between diverging plates. It typically features a series of 200–20 000 km-long segments offset by transform faults.

mineral A naturally occurring solid inorganic element or compound that exhibits a regular internal crystal structure and a distinctive chemical composition. Studies have shown that more than 2000 minerals exist, but about 20 minerals are so abundant that they account for more than 95% of the mineral mass in the Earth's crust. These are known as the rock-forming minerals. The most common rock-forming minerals are silicates, so named because their chief constituents are oxygen (O) and silicon (Si), which bond to form silica ($SiO4$). In turn, the silica bonds with one or more of the elements magnesium (Mg), iron (Fe), calcium (Ca), aluminium (Al), sodium (Na), and potassium (K) to form silicate minerals, such as olivine, pyroxene, amphibole, biotite, plagioclase, alkali feldspar, and muscovite.

móberg lithified tephra and other pyroclastic material that has been subjected to palagonitization (i.e. alteration of the volcanic glass via hydration).

obsidian Dark volcanic glass of felsic composition.

oceanic crust The crust underlying the oceanic basins, mainly composed of basaltic volcanic rocks.

ogives A dark, curved arcuate structure, one of a series repeated periodically down a glacier, typically formed at a base of an icefall, and resembling the pointed arch or rib across a Gothic vault.

partial melting Melting points of igneous minerals change systematically, where olivine and calcium-rich plagioclase have the highest melting temperature and quartz has the lowest. When a rock is heated, the minerals with low melting temperatures dissolve first and those with the highest melting temperatures are most resistant to melting. If all of the minerals that make up a rock are melted, then the magma will have the same bulk composition as the rock. On the other hand, if only the minerals with the lowest melting point are dissolved in the process, then the rock has been subjected to partial melting and will have different composition than the rock from which it is derived. At relatively low temperatures (700–900°C) only the felsic minerals, such as quartz, feldspar, or muscovite are dissolved, and melting produces **felsic** magmas. Conversely, melting at high temperatures (>1100°C) yields melts that are rich in olivine, plagioclase and pyroxene, and thus form **mafic** magmas. Melting at temperatures between these two extremes may form magmas of intermediate compositions.

pyroclast An individual fragment or clast in pyroclastic deposits or rocks formed by disintegration of magma in explosive eruptions.

regolith Any solid material lying on top of bedrock, including soils, alluvium, and rock fragments weathered from the bedrock.

sandur An Icelandic word, it is used internationally to denote broad plains of glaciofluvial deposits formed by braided river systems.

spherulite Radiating aggregates of alkali feldspar, 0.1–2 cm in diameter, formed by crystallization of volcanic glass while the lava is still hot, but at temperatures below the freezing point of the magma. This process is known as devitrification.

stony rhyolite Crystalline felsic lava formed by devitrification of volcanic glass during slow cooling. Typically features fine-grain intergrowths of quartz and alkali feldspar known as felsitic texture.

strandlines An ephemeral line or level at which a body of water (i.e. the sea or a lake) meets the land. Also commonly used to refer to shorelines now elevated above present water level.

syncline A fold with younger layers closer to the centre of the structure.

transform or **transform fault** A strike-slip fault connecting the ends of offset segments in mid-ocean ridges.

tidewater glacier Glaciers that terminate in the sea.

volcanic belt A belt where volcanism is associated with strike-slip or transform faulting.

volcanic succession Volcanic rock units or mass of volcanic strata that succeed one another in chronological order.

volcaniclastic deposit A debris or consolidated clastic rock predominantly composed of volcanic particles.

volcanotectonic Geological structures or events that result from or are controlled by both volcanic and tectonic processes.

Bibliography

Bergh, S. G. & G. E. Sigvaldason 1991. Pleistocene mass-flow deposits of basaltic hyaloclastite on a shallow submarine shelf, South Iceland. *Bulletin of Volcanology* **53**, 597–611.

Bjornsson, H. & P. Einarsson 1990. Volcanoes beneath VatnaJökull, Iceland: Evidence from radio-echo-sounding, earthquakes, and jokulhlaups. *Jökull* **40**, 147–67.

Bjornsson, A., K. Saemundsson, P. Einarsson, E. Tryggvason, K. Gronvold 1977. Current rifting episode in North Iceland. *Nature* **266**(5600), 318–23.

Blake, D. H. 1966. The net-veined complex of the Austurhorn intrusion, Southeast Iceland. *Journal of Geology* **74**, 891–907.

Einarsson, P. & S. Bjornsson 1979. Earthquakes in Iceland. *Jökull* **29**, 37–43.

Einarsson, Þ. 1994. *Geology of Iceland – rocks and landscape.* Reykjavík: Mál og Menning.

Geirsdóttir, Á. & J. Eiríksson 1994a. Growth of an intermittent ice sheet in Iceland during the late Pliocene and early Pleistocene. *Quaternary Research* **42**, 115–30.

Geirsdóttir, Á. & J. Eiríksson 1994b. Sedimentary facies and environmental history of the late-glacial-glaciomarine Fossvogur sediments in Reykjavik, Iceland. *Boreas* **23**, 164–76.

Gudmundsson, A. 1995. Infrastructure and mechanics of volcanic systems in Iceland. *Journal of Volcanology and Geothermal Research* **64**, 1–22.

Gudmundsson, A. 2000. Dynamics of volcanic systems in Iceland: example of tectonism and volcanism at juxtaposed hotspot and mid-ocean ridge systems. *Annual Review of Earth and Planetary Sciences* **28**, 107–140.

Guðmundsson M.T., T. Thordarson, Á. Höskuldsson, G. Larsen, H. Björnsson, F.J. Prata, Oddsson B., Magnússon E., Högnadóttir Th., Petersen G.N., C.L. Hayward, J.A. Stevenson, I. Jónsdóttir, 2012. Ash generation and distribution from the April-May 2010 eruption of Eyjafjallajökull, Iceland. *Scientific Reports*, **2**, 572. DOI: 10.1038/srep 00572.

Gunnarsson, B. 1987. *Petrology and petrogenesis of silicic and intermediate lavas on a propagating oceanic rift: the Torfajökull and Hekla central volcanoes, South-Central Iceland.* PhD thesis, Johns Hopkins University, Baltimore.

Hjartarson, Á. 1991. A revised model of Weichselian deglaciation in South and Southwest Iceland. In *Environmental change in Iceland B past and present*, J. K. Maizels & C. Caseldine (eds), 67–77. Amsterdam: Kluwer.

Hartley M.E. & T. Thordarson, 2012. Formation of Öskjuvatn caldera at Askja, North Iceland: Mechanism of caldera collapse and implications for the lateral flow hypothesis. *Journal of Volcanology and Geothermal Research,* **227-228**: 85–101.

Hartley M.E. & T. Thordarson, 2013. The 1874-76 volcano-tectonic episode at Askja, North Iceland: lateral flow revisited. G3 (Geochemistry, Geophysics, Geosytems); DOI:10.1002/ggge.20151.

Höskuldsson Á, ., K. Vogfjord, N. Oskarsson, R. Petersen, K. Grönvold, 2007.The millennium eruption of Hekla in February 2000. *Bulletin of Volcanology*, **70**, 169–182.

Jakobsson, S.P., 1968. The geology and petrology of the Vestmann Islands. A preliminary report. *Surtsey Research Progress Report*, **4**, 113–29.

Jakobsson, S.P., 1979. Petrology of recent basalts of the Eastern Volcanic Zone, Iceland. *Acta Naturalia Islandica* **26**, 1–103.

Jóhannesson, H. 1980. Jarðlagaskipan og þróun rekbelta á Vesturlandi. *Náttúrufræðingurinn* **50**, 13–31.

Jóhannesson, H. & K. Sæmundsson, 1998. *Geological map of Iceland* (2nd ed.). Reykjavík: Icelandic Institute of Natural History.

Jones, J. G., 1969. Intraglacial volcanoes of the Laugarvatn region, Southwest Iceland, 1. Geological Society of London, Quarterly Journal **124**, 197–211.

Jones, J. G., 1970. Intraglacial volcanoes of the Laugarvatn region, Southwest Iceland, 2. *Journal of Geology* **78**(2), 127–40.

Jónsson, J. 1978. Jarðfræðikort af Reykjanesskaga. I. Skýringar við jarðfræðikort. (Geological map of the Reykjanes Peninsula). National Energy Authority Research Report OS-JHD 7831, Orkustofnun, Reykjavík.

Jónsson, J. 1988. Hestgerðismúli í Suðursveit. *Náttúrufræðingurinn* **58**, 87–96.

Jökull Special issue: the geology of Iceland, volume **29**.

Jökull Special Issue: The dynamic geology of Iceland, volume **58**.

Jude-Eton T, T. Thordarson, M.T. Gudmundsson, B. Oddsson, 2012. Dynamics, stratigraphy and proximal dispersal of supraglacial tephra during the ice-confined 2004 eruption at Grímsvötn Volcano, Iceland. *Bulletin of Volcanology*, DOI 10.1007/s00445-012-0583-3

Kjartansson, G. 1973. Aldur Búrfellshrauns við Hafnarfjörð. *Náttúrufræðingurinn* **42**, 159–83.

Larsen, G. 1984. Recent volcanic history of the Veidivotn fissure swarm, southern Iceland: an approach to volcanic risk assessment. *Journal of Volcanology and Geothermal Research* **22**, 33–58.

Larsen, G. 2000. Holocene eruptions within the Katla volcanic system, South Iceland: characteristics and environmental impact. *Jökull* **49**, 1–28.

Larsen G., 2010 Katla: Tephrochronology and Eruption History. *Developments in Quaternary Sciences*, **13**, 23-48.

Larsen, G. & S. Thorarinsson 1977. H4 and other acid Hekla tephra layers. *Jökull* **27**, 28–46.

Larsen, G., M.T. Gudmundsson, H. Bjornsson 1998. Eight centuries of periodic vol- canism at the center of the Iceland hotspot revealed by glacier tephrostratigraphy. *Geology* **26**, 943–6.

Larsen, G., A. Dugmore, A. Newton 1999. Geochemistry of historical-age silicic tephras in Iceland. *Holocene* **9**, 463–71.

Martin E., J.L. Paquette, V. Bosse, G. Ruffet, M. Tiepolo, O. Sigmarsson, 2011. Geodynamics of rift–plume interaction in Iceland as constrained by new 40Ar/39Ar and in situ U–Pb zircon ages. *Earth and Planetary Science Letters* **311**, 28–38.

Mattson, H. and Á. Höskuldsson 2003. Geology of the Heimaey volcanic centre, south Iceland: early evolution of a central volcano in a propagating rift? *Journal of Volcanology and Geothermal Research* **127**, 55–71.

McPhie, J., M. Doyle, R. Allen 1993. *Volcanic textures: a guide to the interpretation of textures in volcanic rocks*. Hobart: Tasmanian Government Printing Office.

Norddahl, H. & H. Haflidason 1992. The Skógar tephra, a Younger Dryas marker in North Iceland. *Boreas* **21**, 23–41.

Oskarsson, N., G.E. Sigvaldason, S. Steinþorsson 1982. A dynamic model of rift zone petrogenesis and the regional petrology of Iceland. *Journal of Petrology* **23**(1), 28–74.

Sanders, A. D., J.G. Fitton, A.D. Kerr, M.J. Norry, R.W. Kent 1997. North Atlantic Province. In *Large igneous provinces: continental, oceanic, and planetary flood volcanism*, J. J. Mahoney & M. F.Coffin (eds), 44–93. Geophysical Monograph 100, American Geophysical Union, Washington DC.

Self, S., L. Keszthelyi and T. Thordarson 1998. The importance of pahoehoe. Ann. Rev. Earth Planet. Sci. **26**, 81–110.

Sigbjarnason G., 1996 Norðan Vatnajökuls III: Eldstöðar og hraun frá nútíma (North of Vatnajökull III: Volcanic Holocene vents and lava flows). *Nátturfræðingurinn* **65**(3-4), 199-212.

Sigmarsson, O., C. Hemond, M. Condomines, S. Fourcade, N. Oskarsson 1991. Origin of silicic magma in Iceland revealed by Th isotopes. *Geology* **19**, 621–4.

Sigurgeirsson, M. Á. 1995. Yngra Stampagosið á Reykjanesi. *Nátturufræðingurinn* **64**, 211–30.

Steinþórsson, S. 1966. The ankaramites of Hvammsmúli, Eyjafjöll, southern Iceland. *Acta Naturalia Islandica* **2**, 1–32.

Steinþórsson, S. 1981. Ísland og Flekakenningin. In *Náttúra Íslands*, 29–63. Reykjavík: Almenna Bókafélagið.

Steinþorsson, S., N. Oskarsson, G.E. Sigvaldason 1985. Origin of alkali basalts in Iceland: a plate tectonic model. *Journal of Geophysical Research* **90**(B12), 10027–10042.

Sæmundsson, K. 1979. Outline of geology of Iceland. *Jökull* **29**, 7–28.

Sæmundsson, K. 1991. Jarðfræði Kröflukerfisins. In *Náttúra Mývatns*, 25–95. Reykjavík: Hið Íslenska Náttúrufræðifélag.

Sæmundsson, K. 1992. Geology of Thingvallavatn area. *Oikos* **64**, 40–68.

Thorarinsson, S. 1958. The Öraefajökull eruption of 1362. *Acta Naturalia Islandica* **2**, 1–100.

Thorarinsson, S. 1967. *The eruptions of Hekla in historical times: a tephrochronological study* [in the series The Eruption of Hekla 1947–48]. Special Publications 1, Societas Scientiarum Islandica, Reykjavík.

Thorarinsson, S. 1969. Surtsey: the new island in the North Atlantic. London: Cassell.

Thordarson, T. & S. Self 1993. The Laki (Skaftár Fires) and Grímsvötn eruptions in 1783–1785. *Bulletin of Volcanology*, **55**, 233–63.

Thordarson, T. & S. Self 2001. Real-time observations of the Laki sulfuric aerosol cloud in Europe 1783 as documented by Professor S. P. van Swinden at Franeker, Holland. *Jökull* **50**, 65–72.

Thordarson, T., D.J. Miller, G. Larsen, S. Self, H. Sigurdsson 2001. New estimates of sulfur degassing and atmospheric mass-loading by the AD 934 Eldgjá eruption, Iceland. *Journal of Volcanology and Geothermal Research* **108**(1–4), 33–54.

Thordarson T. and Larsen G., 2007. Volcanism in Iceland in Historical Time: Volcano types, eruption styles and eruptive history. *Journal of Geodynamics*, **43**, 1: 118-152.

Thordarson T. and Höskuldsson Á., 2008. Postglacial volcanism in Iceland. *Jökull*, **58**: 197-228.

Thordarson T. & Sigmarsson O, 2009. Effusive activity in the 1963-67 Surtsey eruption, Iceland: flow emplacement and growth of small lava shields In Thordarson et al (eds), *Studies in Volcanology: The Legacy of George Walker*. Spec Publ IAVCEI No **3**; 53-84.

Walker, G.P.L. 1959. Geology of the Reydarfjordur area, eastern Iceland. *Quarterly Journal of Geological Society of London* **114**, 367–91.

Walker, G.P.L. 1960. Zeolite zones and dike distribution in relation to the structure of the basalts of eastern Iceland. *Journal of Geology* **68**, 515–27.

Walker, G.P.L. 1962. Tertiary welded tuffs in Iceland. *Geological Society of London, Quarterly Journal* **118**, 275–93.

Walker, G. P. L. 1963. The Breiðdalur central volcano, eastern Iceland. *Geological Society of London, Quarterly Journal* **119**, 29–63.

Walker, G.P.L. 1964. Geological investigations in eastern Iceland. *Bulletin of Volcanology* **27**, 1–15.

Walker, G.P.L. 1966. Acid volcanic rocks in Iceland. *Bulletin of Volcanology* **29**, 375–402.

Index

Following Icelandic convention, the ð/Ð (edh or eth) character falls between d and e; the þ/Þ character (thorn) follows z. Vowels follow conventional rules of alphabetical sequencing. Page numbers in *italic* denote figures. Page numbers in **bold** denote tables.

GPS coordinates

GPS coordinates for locations/sites mentioned in the text

Location   Page   **Decimal degrees Lat / Long**

Aðaldalur   184   **65.8936 / −17.3849**
Álftafjörður   207   **65.017 / −22.6461**
Álftafjörður   156   **64.5751 / −14.5418**
Álftafjörður   159, 160   **64.5795 / −14.549**
Álftanes   57   **64.1006 / −22.0326**
Álftaver   126,128, 133   **63.5217 / −18.3783**
Almannagjá   82, 84   **64.2563 / −21.1277**
Ármannsfell   81   **64.319 / −21.0339**
Árnes   91   **64.0414 / −20.251**
Ásbyrgi   174   **66.0091 / −16.5061**
Askja   217, 219   **65.0349 / −16.7592**
Austurhorn   159, 160   **64.4122 / −14.5439**
Bæjarfell   68   **63.8155 / −22.7043**
Berserkjahraun   205   **64.9641 / −22.964**
Berufjörður   162   **64.7599 / −14.4369**
Bifröst   214   **64.766 / −21.5538**
Bjarkarlundur   195   **65.5573 / −22.1145**
Bláfjall   218   **65.4422 / −16.844**
Bláfjall   180   **65.4431 / −16.8552**
Bláhnjúkur   234   **63.9771 / −19.0675**
Bleiksmýrardalur   197   **65.405 / −17.7749**
Bólstaðarhlíðarfjall   201   **65.5339 /
    −19.8323**
Borgarfjörður anticline   208   **64.6348 /
    −21.7911**
Borgarhöfn   157   **64.1963 / −15.7805**
Borgarhólar lava shield   80   **64.152 /
    −21.4673**
Borgarnes   214   **64.5531 / −21.9003**
Breiðamerkurjökull   152, 156   **64.0546 /
    −16.4607**
Breiðamerkursandur   151   **64.0391 /
    −16.2681**
Breiðavík   175   **66.1868 / −17.1564**
Breiðdalsvík   164   **64.7762 / −14.0303**
Breiðdalur   164   **64.8122 / −14.1913**
Brimurð   104   **63.4069 / −20.2747**
Brjánslækur   195   **65.5261 / −23.1924**
Brókárjökull   157   **64.2578 / −16.139**
Bruni   65   **64.0298 / −21.9707**
Búðafoss   91   **64.0328 / −20.3294**
Búlandshöfði   210   **64.9429 / −23.4889**
Búlandshöfði   205   **64.9425 / −23.4759**
Búlandssel   123   **63.7833 / −18.5833**
Búrfell   62–4   **64.0333 / −21.8303**
Búrfell   180   **65.5543 / −16.6466**
Búrfell   218   **65.5522 / −16.6504**

Búrfell Lava Site 1 – inflated
    margin   64   **64.0751 / −21.9041**
Búrfell Lava Site 1 – lava tube   64   **64.0717 /
    −21.8926**
Búrfell Power Plant   95   **64.1052 / −19.8332**
Dagmálafell   83   64.1766 / −21.0546
Dettifoss   174   **65.8142 / −16.3846**
Dimmuborgir   184, 187   **65.5915 / −16.9096**
Dómadalshraun   235   **64.0297 / −19.1009**
Dverghamrar   138   **63.8506 / −17.8602**
Dyrhólaey   120   **63.4076 / −19.1138**
Dysjarhóll   118   **63.5745 / −19.8754**
Eiríksjökull   203   **64.7734 / −20.3964**
Eldborg á Mýrum   206   **64.7956 / −22.3208**
Eldborgir   82   **64.2389 / −20.9371**
Eldfell   103, 105–6   **63.4323 / −20.2497**
Eldfell lava, Kirkjubæjarhraun   106   **63.4393 /
    −20.243**
Eldfell lava, Páskahraun   106   **63.43 /
    −20.2402**
Eldfell lava, site 2   106   **63.4424 / −20.2599**
Eldfell lava, site 3   106   **63.441 / −20.2646**
Eldfell lava, site 4   106   **63.4434 / −20.2608**
Eldfell lava, site 5   106   **63.4365 / −20.2278**
Eldgjá   128   **63.9629 / −18.6157**
Eldri Stampar   72   **63.8251 / −22.7173**
Eldsveitirnar   109   **63.6845 / −18.2833**
Engimýri   199   **65.5716 / −18.5351**
Esja   51–52   **64.2407 / −21.6659**
Eyjafjallajökull   109, 111   **63.6297 /
    −19.6369**
Eyjafjallajökull volcano   79   **63.6276 /
    −19.636**
Eyrarfjall   51, 52   **64.3193 / −21.7263**
Fagradalsfjall   73   **63.8962 / −22.2934**
Fellsfjall   152   **64.1294 / −16.1367**
Festarfjall   73   **63.8573 / −22.3373**
Figure 2.1 Site 4   50   **64.048 / −21.9462**
Figure 2.1 Site 5   50   **64.0475 / −21.8491**
Figure 2.1 Site 6   50   **64.1471 / −21.7216**
Fimmvörðuháls   114   **63.626 / −19.4443**
Fjaðrárgljúfur   131,138   **63.773 / −18.1745**
Fjallsárjökull   152   **64.0163 / −16.4056**
Flatadyngja   219   **65.1346 / −16.4666**
Fljótsdalshérað   162   **65.2256 / −14.533**
Fljótshlíð   100   **63.7349 / −20.0214**
Fljótshverfi   128   **63.9142 / −17.717**

261

Fnjóskadalur 197 65.7516 / -17.8839
Fossá 52, 94 64.3561 / -21.4559
Fossvogur 53 64.1213 / -21.9264
Furuvík 175 66.1763 / -17.2452
Garðar 207 64.8038 / -22.2608
Garðskagi 67 64.0811 / -22.6895
Gaukshöfði 92 64.0778 / -20.0245
Geysir 79, 86–88 64.3104 / -20.3031
Gígjökull 115 63.6628 / -19.6212
Gígöldur 219 64.8817 / -16.9405
Gjábakki 86 64.1969 / -21.0275
Gjain 94 64.1514 / -19.7382
Grábrók 205 64.7701 / -21.5376
Grænalón 141 64.1688 / -17.3605
Grænavatn 75 63.8848 / -22.0537
Grænavatnsbruni 184 65.506 / -16.9633
Grafarlandaá 222 65.3367 / -16.0702
Grimsnes 90–1 64.042 / -20.8867
Grímsvötn 45, 128 64.4035 / -17.3417
Grindavík 72 63.8453 / -22.4351
Grjotargljufur 93 64.1884 / -19.8839
Grundarfjörður 210 64.9242 / -23.2402
Gullfoss 79, 88 64.3255 / -20.1269
Gýgjukvísl–Skeiðará 140 63.9394 / -17.3656
Hæðir lava shield 80 64.1731 / -21.2703
Hafnarfjall 208 64.4834 / -21.9042
Hafnir 67 63.934 / -22.6838
Hafursey 123 63.519 / -18.7739
Hágöng 182 65.74 / -16.6861
Háin 103 63.4404 / -20.2873
Háleyjarbunga 72 63.8167 / -22.651
Hallbjarnarstaðará 178 66.1417 / -17.2425
Heimaey 102, 103–4 63.4427 / -20.2751
Hekla 95 64.0133 / -19.5922
Hekla volcano 79 63.9951 / -19.6505
Helgafell 104 63.4292 / -20.2601
Hellnahraun 65 64.0376/ -21.9769
Hengill 83 64.0772 / -21.3132
Herðubreið 219 65.1753 / -16.3464
Herðubreiðarlindir 219, 222 65.1964 / -16.2239
Herðubreiðartögl 219 65.1036 / -16.3544
Hestfjall 88 64.0127 / -20.6687
Hítardalur 208 64.8279 / -22.0651
Hítarvatn 208 64.883 / -21.9605
Hjallar 62–63 64.0502 / -21.8453
Hjalparfoss 94 64.1145 / -19.8536
Hjörleifshöfði 120 63.4166 / -18.7552
Hliðarfjall 182 65.6775 / -16.8566
Hljóðaklettar 175 65.9464 / -16.533
Hnausahraun 238 64.0683 / -19.0453
Höfðabrekka 122 63.4267 / -18.8982
Hofsá River 118 63.5238 / -19.4552
Hofsjökull 79 64.8135 / -18.8063
Hólafjall 199 65.5345 / -20.7064
Hólmsárbrú 129 63.6354 / -18.5199

Holtavörðuheiði 203 64.9656 / -21.0622
Holuhraun 219 64.8426 / -16.8452
Hornafjörður 159 64.2535 / -15.2089
Hornarfjarðarfljót 157 64.355 / -15.367
Höskuldsvík 178 66.1569 / -17.2781
Hrafnabjörg 82 64.2723 / -20.9272
Hrafnagjá 84 64.2573 / -21.1108
Hrafntinnuhraun 235 63.9527 / -19.2744
Hrafntinnusker 238 63.9338 / -19.1891
Hreðavatn 208 64.7588 / -21.5813
Hrólfsvík 73 63.8495 / -22.3662
Hrómundartindur 83 64.0762 / -21.2016
Hrossaborg 218 65.613 / -16.2617
Hrutagja lava shield 66–7 64.0341 / -22.0979
Hrutagja, Rauðimelur quarry 66 64.0317 / -22.0653
Hrútárjökull 152 64.0072 / -16.4556
Húsavík 180 66.0446 / -17.3394
Húsavíkurfjall 180 66.0471 / -17.3031
Húsavíkurkleif 195 65.643 / -21.6342
Hvalfell 215 64.3858 / -21.2125
Hvammsmúli 117 63.5759 / -19.8715
Hvammsvík 52 64.3677 / -21.5679
Hvamsfjörður 207 65.0747 / -22.0632
Hvannalindir 233 64.8693 / -16.3267
Hverfjall 185 65.6052 / -16.8772
Jarðbaðshólar 186 65.6319 / -16.8601
Jökulsá 152 64.0438 / -16.1794
Jökulsá a Solheimasandi 119 63.4984 / -19.3989
Jökulsárgljúfur 171 65.8213 / -16.3849
Jökulsárlón 152 64.0537 / -16.1806
Kaldakvísl 178 66.1017 / -17.2757
Kálfafellsdalur 157 64.1868 / -15.9489
Kálfstindar 86 64.2521 / -20.8576
Kapelluhraun 65 64.0246 / -21.9697
Katla, see Myrdalsjokull/Katla
Kárastaðir 162 64.6972 / -14.2271
Kerhóll 90 64.0583 / -20.8445
Kerið 90 64.0412 / -20.8851
Ketildyngja 184 65.4293 / -16.6544
Ketildyngja 218 65.4466 / -16.6558
Ketillaugarfjall 159 63.3422 / -15.2272
Kirkjubæjarklaustur 133,138 63.7895 / -18.0528
Kirkjufell 234 63.9841 / -18.9443
Klauf 104 63.4105 / -20.2839
Klauf, site 1 104 63.4076 / -20.2787
Kleifarvatn 75 63.9264 / -21.9745
Kögur 196 65.9141 / -23.8316
Kolbeinsey 180 67.117 / -18.6
Kolgríma 157 64.2463 / -15.6743
Kollóttadyngja 219 65.2186 / -16.5531
Kópavogur inlet 56 64.1051 / -21.9066
Kotá 200 65.4424 / -19.0367
Kotagil 201 65.4544 / -19.0569

Kötlugjá   122   **63.6076 / -19.022**
Kötlutangi   125   **63.3986 / -18.7456**
Kræðuborgir   218   **65.6224 / -16.5682**
Krafla   182   **65.7149 / -16.729**
Kriki   123   **63.6238 / -18.8682**
Krýsuvík   73   **63.8869 / -22.0652**
Kverkfjöll   217, 231   **64.6874 / -16.6028**
Kverkhnjúkar   231   **64.7683 / -16.4937**
Lake Mývatn   180   **65.6006 / -16.9944**
Laki   128   **64.0701 / -18.2388**
Lambafit   99   **64.074 / -19.4043**
Landbrotshólar   57, 134   **63.7597 /
-17.9708**
Landmannalaugar   217   **63.989 / -19.1797**
Landmannaleið   95   **64.0923 / -19.7476**
Langavatn   207   **64.7829 / -21.7842**
Langidalur   201   **65.5736 / -19.9996**
Langjökull   79, 84   **64.6718 / -20.0869**
Laufafell   234   **63.9155 / -19.3415**
Laugahraun   233   **63.9881 / -19.0736**
Laugarvatn   86   **64.2178 / -20.732**
Laugarvatnsfjall   86   **64.2296 / -20.7738**
Laxárdalur   184   **65.6362 / -17.1636**
Lindá   222   **65.2296 / -16.1852**
Lindahraun   233   **64.8513 / -16.3196**
Ljósufjöll   205   **64.9146 / -22.5803**
Lögurinn   169-70   **65.1675 / -14.6501**
Lokufjall   51   **64.2818 / -21.826**
Lómagnúpur   128, 138   **63.9639 / -17.5036**
Lón   159   **64.4085 / -14.662**
Lúdent   182   **65.5814 / -16.8132**
Lúdentsborgir   184   **65.5775 / -16.8347**
Lyngdalsheiði   86   **64.1536 / -20.8887**
Lýsuskarð   205   **64.8521 / -23.2075**
Mælifell   160, 162   **64.4723 / -14.5109**
Markarfljót   100   **63.6175 / -20.0156**
Markarfljót bridge   111, 115   **63.6164 /
-20.0152**
Miðfell   83   **64.1783 / -21.0519**
Miðnes   67   **63.997 / -22.6463**
Mókollsdalur   195-196   **65.5214 / -21.5302**
Móskarðshnjúkar   51, 80   **64.2429 / -21.5283**
Mt Búrfell   94   **64.0816 / -19.812**
Mt Fossalda   93   **64.2092 / -19.7491**
Mt Ingólfsfjall   90   **63.9765 / -21.0252**
Mt Rauðufossafjöll   100   **63.9957 / -19.3762**
Mt Reykholt   93   **64.1446 / -19.841**
Mt Skeljafell   93, 94   **64.1289 / -19.8025**
Mundafell   99   **63.9845 / -19.5635**
Mýrar   213   **64.6783 / -22.0753**
Mýrdalsjökull volcano   79, 122-7   **63.6365 /
-19.1164**
Mýrdalsjokull/Katla   109, 111,
122-7   **63.6329 / -19.0546**
Mýrdalssandur   111, 126-7   **63.4796 /
-18.6828**
Mýrdalur   111   **63.4472 / -19.112**

Námafjall   218   **65.6391 / -16.8205**
Námshraun   235   **63.9999 / -19.0484**
Nesjahraun   83   **64.1394 / -21.2137**
Nesjavellir   83   **64.1083 / -21.2569**
Norðurárdalur   200   **65.4457 / -19.007**
Norðurklettar   103   **63.4488 / -20.2645**
Núpsvötn-Súla   140   **63.9559 / -17.4671**
Ódáðahraun   218   **65.1804 / -16.7716**
Ogmundarhraun   65, 73-75   **63.836 /
-22.1655**
Ólafsvík   209   **64.8947 / -23.706**
Ölkelduháls   83   **64.0594 / -21.2338**
Öræfajökull   109, 128-9   **63.9886 / -16.6466**
Öræfi   109   **63.9611 / -16.8674**
Óseyrarbrú   79   **63.8787 / -21.2124**
Öxarfjörður   182   **66.217 / -16.7369**
Öxnadalur   199   **65.5873 / -18.5301**
Pétursey   120   **63.4673 / -19.2711**
pumice fall - H3 tephra outcrop   94   **64.1283
/ -19.8247**
Randhólar   174   **65.8491 / -16.3781**
Rauðhálsar   213   **64.7463 / -21.8842**
Rauðhólar   57-61   **64.0929 / -21.7491**
Rauðimelur (spit bar)   67   **63.9286 /
-22.4809**
Rauðimelur Quarry   66   **64.0316 / -22.0653**
Rauðubjallar   99   **63.9425 / -19.7529**
Rauðuborgir   174   **65.6012 / -16.4743**
Rauðufossafjöll   234   **63.9961 / -19.3706**
Reká   178   **66.1146 / -17.2505**
Reykjahlíð   182   **65.6449 / -16.9059**
Reykjanes   68-9   **63.8129 / -22.7148**
Reykjanesviti/Bæjarfell   67   **63.8155 /
-22.7043**
Reykjavik   49   **64.1328 / -21.8983**
Road 32   91   **64.0461 / -20.4116**
Röndólfur   164   **64.8023 / -14.3703**
Sæfell   104   **63.4159 / -20.275**
Sandey   83   **64.1838 / -21.1654**
Sandfell   166   **64.8766 / -13.8972**
Sandfellshæð   66-7, 72   **63.8603 / -22.5748**
Saurbær   195   **65.4774 / -24.0025**
Selárdalur   195   **66.1226 / -23.4395**
Seljalandsfoss   99, 100   **63.6162 / -19.994**
Sellandafjall   182   **65.4088 / -17.0392**
Seyðishólar   90   **64.0658 / -20.8426**
Síða   128   **63.8395 / -17.9661**
Skaftafell   111, 147   **64.027 / -16.995**
Skaftártunga   127   **63.6939 / -18.4989**
Skagi   195   **65.8937 / -20.0581**
Skálafell   72   **63.8125 / -22.6822**
Skammidalur   120   **63.4513 / -19.1006**
Skarfatangi   104   **63.4229 / -20.2522**
Skeiðará River ~Bridge   140-141   **63.9758/
-17.0003**
Skeiðarárjökull   140-141   **64.0182 /
-17.2439**

Skeiðarársandur   109, 111, 140, 141   **63.9486 / -17.4128**
Skessa   168   **65.0082 / -14.276**
Skjaldbreiður   78, 81-2   **64.4089 / -20.7522**
Skógafoss   118   **63.5312 / -19.5124**
Skogar   118   **63.5279 / -19.4995**
Skogasandur   111, 119   **63.504 / -19.4832**
Skorradalsvatn   214   **64.5172 / -21.4851**
Skriðuá   168   **64.8105 / -14.3335**
Slaufrudalur   160   **64.3239 / -15.0081**
Slúttnes   183   **65.6419 / -16.9553**
Snæfellsjökull   212, 210   **64.806 / -23.7769**
Sog   90   **64.0049 / -20.9731**
Solheimar   118   **63.4947 / -19.3282**
Solheimasandur   111, 118   **63.4735 / -19.3761**
Stapafell   83   **64.0922 / -21.1723**
Stapafell   67   **63.9055 / -22.5248**
Stardalur central volcano   80   **64.2203 / -21.5411**
Steinavötn   157   **64.1655 / -15.9765**
Stöð   210   **64.9562 / -23.3659**
Stöng   94   **64.1517 / -19.7524**
Stórhöfði   103   **63.4007 / -20.2869**
Sultartangi   95   **64.1807 / -19.5613**
Surtsey   101, 107-8   **63.3015 / -20.6039**
Svartadyngja   219   **65.1076 / -16.5278**
Svartshengi   72   **63.8801 / -22.432**
Sveinaborgir   218   **65.6251 / -16.404**
Sveinagjá   218   **65.441   16.4822**
Sveinar   174   **65.6858 / -16.4588**
Svínahraunsbruni   91   **64.0326 / -21.4616**
Sýrfell   68   **63.8369 / -22.6599**
Tindaskagi   82   **64.3354 / -20.8166**
Tindfjöll volcano   79   **63.7953 / -19.5908**
Tinnudalur   169   **64.8717 / -14.1818**
Tjaldastaðargjá   72   **63.8529 / -22.6554**
Tjarnahnúkur   83   **64.065 / -21.2158**
Tjörnes   175   **66.1519 / -17.0866**

Tjörvahraun   238   **64.1356 / -19.1511**
Torfajökull   234   **63.9014 / -19.0221**
Torfajökull volcano   79   **63.9487 / -19.1399**
Tvibollahraun   65   **64.0376 / -21.9769**
Valahnúkur   68, 70   **63.8109 / -22.7115**
Vatnajökull   109   **64.3943 / -16.978**
Vatnaöldur   238   **64.1175 / -18.9173**
Vatnsdalsfjall   201   **65.4592 / -20.2493**
Vatnsdalshólar   201   **65.4956 / -20.3779**
Vatnsfell   69, 70, 71   **63.815 / -22.7234**
Vatnsnes   202   **65.5867 / -20.7963**
Veiðivötn   217, 235   **64.142 / -18.7663**
Vesturhorn   159   **64.2768 / -14.9549**
Viðborðsfjall   159   **64.3438 / -15.4172**
Vilhjálmsvík   104   **63.4112 / -20.2865**
Víti   225-6   **65.0462 / -16.7263**
Vogar   187   **65.6224 / -16.9203**
Vogarstapi   67   **63.9712 / -22.4139**
Vörðufell   88   **64.0686 / -20.5428**
Yngri Stampar   68-9   **63.8198 / -22.7221**
Ytri-Dalbær   133   **63.7683 / -18.1215**
Þeistareykir   180   **65.1817 / -16.9758**
Þingvellir   79, 80-6   **64.2562 / -21.1285**
Þjófahraun   82-83   **64.3048 / -20.8409**
Þjórsá   91   **64.0603 / -20.0755**
Þjórsá Bridge   99   **63.9298 / -20.6649**
Þjórsá lava (inbetween Selfoss and Þjórsá bridge)   77   **63.9508 / -20.6928**
Þjórsárdalur   80, 91, 93-4   **64.1217 / -19.8713**
Þorbjarnarfell   72   **63.8642 / -22.4405**
Þórðarfell   67   **63.891 / -22.5215**
Þórðarhyrna   141   **64.2685 / -17.5458**
Þorvaldshraun   219   **64.9687 / -16.714**
Þráinskjöldur lava shield   66-7   **63.9618 / -22.3535**
Þrengslaborgir   184, 187   **65.5396 / -16.8367**